Panagiotis D. Christofides · Jinfeng Liu ·
David Muñoz de la Peña

T0075861

Networked and Distributed Predictive Control

Methods and Nonlinear Process Network Applications

 Springer

Panagiotis D. Christofides
Department of Chemical and Biomolecular
Engineering
University of California, Los Angeles
Los Angeles
USA
pdc@seas.ucla.edu

David Muñoz de la Peña
Departamento de Ingeniería de Sistemas
y Automática
Universidad de Sevilla
Camino de los Descubrimientos
41092 Sevilla
Spain
davidmps@cartuja.us.es

Jinfeng Liu
Department of Chemical and Biomolecular
Engineering
University of California, Los Angeles
Los Angeles
USA
jinfeng@ucla.edu

ISSN 1430-9491
ISBN 978-1-4471-2648-5   ISBN 978-0-85729-582-8 (eBook)
DOI 10.1007/978-0-85729-582-8
Springer London Dordrecht Heidelberg New York

British Library Cataloguing in Publication Data
A catalogue record for this book is available from the British Library

Cover design: VTeX UAB, Lithuania

Printed on acid-free paper

Springer is part of Springer Science+Business Media (www.springer.com)

Advances in Industrial Control

Series Editors

Professor Michael J. Grimble, Professor of Industrial Systems and Director
Professor Michael A. Johnson, Professor (Emeritus) of Control Systems and Deputy Director

Industrial Control Centre
Department of Electronic and Electrical Engineering
University of Strathclyde
Graham Hills Building
50 George Street
Glasgow Gl 1QE
UK

Series Advisory Board

Professor E.F. Camacho
Escuela Superior de Ingenieros
Universidad de Sevilla
Camino de los Descubrimientos s/n
41092 Sevilla
Spain

Professor S. Engell
Lehrstuhl für Anlagensteuerungstechnik
Fachbereich Chemietechnik
Universität Dortmund
44221 Dortmund
Germany

Professor G. Goodwin
Department of Electrical and Computer Engineering
The University of Newcastle
Callaghan NSW 2308
Australia

Professor T.J. Harris
Department of Chemical Engineering
Queen's University
Kingston, Ontario
K7L 3N6
Canada

Professor T.H. Lee
Department of Electrical and Computer Engineering
National University of Singapore
4 Engineering Drive 3
Singapore 117576
Singapore

Professor (Emeritus) O.P. Malik
Department of Electrical and Computer Engineering
University of Calgary
2500, University Drive, NW
Calgary, Alberta
T2N 1N4
Canada

Professor K.-F. Man
Electronic Engineering Department
City University of Hong Kong
Tat Chee Avenue
Kowloon
Hong Kong

Professor G. Olsson
Department of Industrial Electrical Engineering and Automation
Lund Institute of Technology
Box 118
221 00 Lund
Sweden

Professor A. Ray
Department of Mechanical Engineering
Pennsylvania State University
0329 Reber Building
University Park
PA 16802
USA

Professor D.E. Seborg
Chemical Engineering
University of California Santa Barbara
3335 Engineering II
Santa Barbara
CA 93106
USA

Doctor K.K. Tan
Department of Electrical and Computer Engineering
National University of Singapore
4 Engineering Drive 3
Singapore 117576
Singapore

Professor I. Yamamoto
Department of Mechanical Systems and Environmental Engineering
Faculty of Environmental Engineering
The University of Kitakyushu
1-1, Hibikino, Wakamatsu-ku, Kitakyushu, Fukuoka, 808-0135
Japan

Series Editors' Foreword

The series *Advances in Industrial Control* aims to report and encourage technology transfer in control engineering. The rapid development of control technology has an impact on all areas of the control discipline. New theory, new controllers, actuators, sensors, new industrial processes, computer methods, new applications, new philosophies..., new challenges. Much of this development work resides in industrial reports, feasibility study papers and the reports of advanced collaborative projects. The series offers an opportunity for researchers to present an extended exposition of such new work in all aspects of industrial control for wider and rapid dissemination.

In some *Advances in Industrial Control* monographs, the author's perspective is one of looking back at successful developments that have found application in practice. Other monographs in the series explore future possibilities, presenting a coherent body of theory with supporting illustrative examples and case studies. This entry to the *Advances in Industrial Control* series, *Networked and Distributed Predictive Control: Methods and Nonlinear Process Network Applications* by Panagiotis D. Christofides, Jinfeng Liu, and David Muñoz de la Peña is a very persuasive exemplar of the "future possibilities" monograph category.

The starting point for the authors' development is the question: if a process has an existing point-to-point (hard-wired) control system, how do we design a networked control system (wired or, more in tune with recent technological developments, wireless) to augment the existing control and what performance benefits can be achieved? What follows from this is a thorough analysis and assessment of different control architectures blended with advanced control design methods. The control design techniques are selected as model predictive control for nonlinear processes but accommodating typical disruptive network characteristics of asynchronous feedback and communication delays.

The reader, whether an industrial engineer or academic researcher, will find a coherent theoretical development that unites model predictive control and Lyapunov stability methods as a control technique termed Lyapunov-based model predictive control. This is shown to have some nice properties of practical utility concerning closed loop stability and the stability region. The authors use this technique and progress through a sequence of increasingly advanced networked control system configurations, devoting a chapter to each particular control structure.

A major strength of the monograph is the attention given to careful and detailed process control examples and case studies that illustrate the characteristics and performance potential of individual networked control systems. One of these is an in-depth case study treatment of a wind–solar energy generation plant, whilst other examples are taken from the chemical process industries. All that is missing from these studies is an estimate of implementation costs and a cost benefit analysis! Process, chemical, and control engineers will find these simulated examples illuminating.

As a forward-looking monograph series on control design, technology, implementation and industrial practice, we are pleased to add this volume to the series as its first entry on networked control systems. As wireless control technology gains in reliability we expect to see many further theoretical and practical developments in this field. This monograph also complements the *Advances in Industrial Control* series's first entry on the closely related field of control using the Internet, so that readers may find the monograph, *Internet-based Control Systems: Design and Applications* (ISBN 978-1-84996-358-9) by Shuang-Hua Yang of interest.

Industrial Control Centre   M.J. Grimble
Glasgow   M.A. Johnson
Scotland, UK

Preface

Traditionally, process control systems rely on control architectures utilizing dedicated, wired links to measurement sensors and control actuators to regulate appropriate process variables at desired values. While this paradigm to process control has been successful, we are currently witnessing an augmentation of the existing, dedicated control systems, with additional networked (wired and/or wireless) actuator/sensor devices which have become cheap and easy-to-install. Such an augmentation in sensor information, actuation capability and network-based availability of data has the potential to dramatically improve the ability of process control systems to optimize closed-loop performance and prevent or deal with abnormal situations more effectively. However, augmenting dedicated control systems with real-time sensor and actuator networks poses a number of new challenges in control system design that cannot be addressed with traditional process control methods, including: (a) the handling of additional, potentially asynchronous and delayed measurements in the overall networked control system, and (b) the substantial increase in the number of process state variables, manipulated inputs and measurements which may impede the ability of centralized control systems (particularly when nonlinear constrained optimization-based control systems like model predictive control are used), to carry out real-time calculations within the limits set by process dynamics and operating conditions.

This book presents rigorous, yet practical, methods for the design of networked and distributed predictive control systems for chemical processes described by nonlinear dynamic models. Beginning with an introduction to the motivation and objectives of this book, the design of model predictive control systems via Lyapunov-based control techniques accounting for networked control-relevant issues, like handling of asynchronous and delayed measurements, is first presented. Then, the book focuses on the development of a two-tier networked control architecture which naturally augments dedicated control systems with networked control systems to maintain closed-loop stability and significantly improve closed-loop performance. Subsequently, the book focuses on the design of distributed predictive control systems, that utilize a fraction of the time required by the respective centralized control systems, to cooperate in an efficient fashion and to compute optimal manipulated input trajectories that achieve the desired stability, performance, and robustness for large-scale nonlinear process networks. Throughout the book, the control methods are applied to large-scale nonlinear process networks and wind–solar energy generation systems and their effectiveness and performance are evaluated through detailed computer simulations.

The book requires basic knowledge of differential equations, linear and nonlinear control theory, and optimization methods and is intended for researchers, graduate students, and process control engineers. Throughout the book, practical implementation issues are discussed to help engineers and researchers understand the application of the methods in greater depth.

In addition to our work, Prof. James F. Davis, Dr. Benjamin J. Ohran, doctoral candidates Mohsen Heidarinejad and Xianzhong Chen, and doctoral student Wei Qi, all at UCLA, contributed substantially to the research results included in the book and in the preparation of the final manuscript. We would like to thank them for their hard work and contributions. We would also like to thank all the other people who contributed in some way to this project. In particular, we would like to thank our colleagues at UCLA and the Universidad de Sevilla for creating a pleasant working environment, and the United States National Science Foundation and the European Commission for financial support. Last, but not least, we would like to express our deepest gratitude to our families for their dedication, encouragement, and support over the course of this project. We dedicate this book to them.

Los Angeles, CA, USA   Panagiotis D. Christofides
Seville, Spain   Jinfeng Liu
                                                                        David Muñoz de la Peña

Contents

Abbreviations

CSTR Continuous stirred tank reactor
DMPC Distributed model predictive control
LCS Local control system
LMPC Lyapunov-based model predictive control
MPC Model predictive control
NCS Networked control system
PI Proportional-integral
PID Proportional-integral-derivative
RHC Receding horizon control

List of Figures

List of Tables

Chapter 1
Introduction

1.1 Motivation

Increasingly faced with the requirements of safety, environmental sustainability, and profitability, chemical process operation is relying extensively on highly automated control systems. This realization has motivated extensive research, over the last forty years, on the development of advanced operation and control strategies to achieve economically optimal plant operation by regulating process variables at appropriate values. With respect to process control, control systems traditionally utilize dedicated, point-to-point wired communication links using a small number of sensors and actuators to regulate appropriate process variables at desired values. While this paradigm to process control has been successful, chemical plant operation could substantially benefit [12, 16, 66, 80, 114, 119] from an efficient integration of the existing, point-to-point control networks (wired connections from each actuator or sensor to the control system using dedicated local area networks) with additional networked (wired or wireless) actuator or sensor devices that have become cheap and easy-to-install. Such an augmentation in sensor information, actuation capability and network-based availability of wired and wireless data is now well underway in the process industries and clearly has the potential to dramatically improve the ability of the single-process and plant-wide model-based control systems to optimize process and plant performance. Network-based communication allows for easy modification of the control strategy by rerouting signals, having redundant systems that can be activated automatically when component failure occurs, and in general, it allows having a high-level supervisory control over the entire plant. However, augmenting existing control networks with real-time wired or wireless sensor and actuator networks challenges many of the assumptions made in the development of traditional process control methods dealing with dynamical systems linked through ideal channels with flawless, continuous communication. In the context of networked control systems, key issues that need to be carefully handled at the control system design level include data losses due to field interference and time-delays due to network traffic as well as due to the potentially heterogeneous nature of the additional measurements. In the context of control system architectures, augmenting dedicated, local control systems with control systems that utilize real-time

P.D. Christofides et al., *Networked and Distributed Predictive Control*,
Advances in Industrial Control,
DOI 10.1007/978-0-85729-582-8_1, © Springer-Verlag London Limited 2011

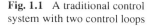

Fig. 1.1 A traditional control
system with two control loops

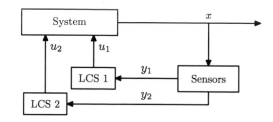

sensor and actuator networks gives rise to the need to coordinate separate control systems that operate on a process. However, the rigorous design of cooperative, distributed control architectures for nonlinear processes is a challenging task that cannot be addressed with traditional process control methods dealing with the design of centralized control systems. To design cooperative, distributed control systems, key fundamental issues that need to be addressed include the design of the individual control systems and of their communication strategy so that they efficiently cooperate in achieving the closed-loop plant objectives. Motivated by the above, this book presents general methods for the design of networked and distributed predictive control systems, accompanied by their application to nonlinear process networks.

1.2 Networked and Distributed Control Architectures

To provide concrete motivation for the control problems addressed in this book, we discuss below the general concept of networked and distributed process control using block diagrams and a chemical process example.

1.2.1 Networked Control Architectures

Traditionally, the different components (i.e., sensor, controller, and actuator) in a control system are connected via wired, point-to-point links, and the control laws are designed and operate based on local continuously-sampled process output measurements. For a system with multiple control loops, the controllers, in general, are designed to work in a decentralized fashion. Figure 1.1 shows a traditional control system with two control loops. In Fig. 1.1, two local control systems (i.e., LCS 1 and LCS 2) are designed based on two different continuously-sampled outputs, y_1 and y_2, of the system. The two controllers do not exchange information and operate in a decentralized fashion.

Communication networks make the transmission of data much easier and provide a higher degree of freedom in the configuration of control systems. However, new issues arise in the design of a networked control system (NCS), for example, the introduction of data losses and time-varying delays in the control loop as well as the use of asynchronous measurements. On the other hand, additional information of

Fig. 1.2 A networked configuration for the system shown in Fig. 1.1 (*dashed lines* denote measurements and control actions transmitted via real-time communication networks)

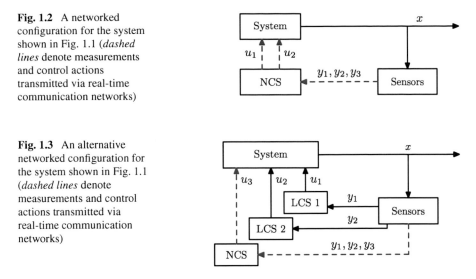

Fig. 1.3 An alternative networked configuration for the system shown in Fig. 1.1 (*dashed lines* denote measurements and control actions transmitted via real-time communication networks)

a system which previously were difficult or impossible to access because of physical or economical reasons may be now available via networked devices like, for example, networked sensors deployed over chemical plants. The additional information may be used to improve the closed-loop performance and the fault tolerance of a control system. However, because of the nature of the additional sensing (for example, concentration versus temperature measurements) and the fact that this information is collected and transmitted through real-time wired or wireless networks, a control system should also be able to handle heterogeneous (for example, continuous, asynchronous and delayed) measurements. In order to take advantages of the use of networks in the transmission of information and to use the additional information provided by networked devices, one approach is to design an NCS which takes data losses, delays and heterogeneous measurements explicitly into account to replace the local control loops. Figure 1.2 shows this kind of NCS design for the system shown in Fig. 1.1. In Fig. 1.2, an NCS is designed to replace the two local controllers in Fig. 1.1 taking into account all the available measurements (i.e., originally available measurements y_1, y_2, and additional measurement y_3). The key issues in the design of such an NCS include the handling of data losses, time-varying delays, and the utilization of heterogeneous measurements.

Instead of replacing the local control loops, an alternative to the above networked control configuration is to design an NCS to augment the local control loops to take advantage of the additional measurements to manipulate additional control inputs or adjust the control actions of the existing local controllers to improve the closed-loop performance. The networked control configuration resulting in this case is shown in Fig. 1.3. The main question is how to design the NCS to maintain the closed-loop stability achieved by the local controllers while improving the closed-loop performance.

1.2.2 Cooperative, Distributed Control Architectures

Consider the second networked control configuration shown in Fig. 1.3. In this configuration, there is no communication between the networked controller and the two
local controllers. In this sense, the three controllers work in a decentralized fashion.
When the local controllers are designed via classical (e.g., proportional-integral-
derivative (PID) control), geometric or Lyapunov-based control methods for which
an explicit formula for the calculation of the control action is available, and the
networked controller is designed via model-based control methods, like model predictive control (MPC), the coupling between the networked controller and the local
controllers may be taken into account if the networked controller is carefully designed. However, when the local controllers are designed via MPC for which there
is no explicit controller formula to calculate the future control actions, it is necessary to establish some, preferably small, communication between the different
controllers so that they can coordinate their actions, which leads to the design of
distributed control systems.

Figure 1.4 shows such a control configuration for the system shown in Fig. 1.1.
In this distributed control system, an LCS is designed to determine u_1 and u_2 and an
NCS is designed to calculate u_3 based on all the information available via networks.
In order to coordinate the control actions, the two controllers communicate to exchange information which could be future input trajectories the two controllers will
apply or/and system measurements. In this case, we need to consider how the distributed controllers should communicate, what information they need to exchange
and how to coordinate their actions to achieve stability of the entire closed-loop
system.

In the distributed control configuration shown in Fig. 1.4, the control inputs are
distributed into the two controllers by their functionalities; that is, the LCS determines u_1 and u_2 to ensure the closed-loop stability, and the NCS determines u_3 to
improve the closed-loop performance. An alternative to this kind of decomposition
of the control inputs is to decompose the inputs spatially; that is, a distributed controller is designed for each control input (or each subsystem) as shown in Fig. 1.5.
In the distributed control configuration of Fig. 1.5, three NCSs are designed to manipulate the three control inputs, respectively, based on all the available measurements. The three controllers communicate to coordinate their actions. This type of
distributed control configuration is more flexible in the control loop selection compared with the one shown in Fig. 1.4.

1.2.3 A Reactor–Separator Process Example

Consider a three vessel, reactor–separator process consisting of two continuously
stirred tank reactors (CSTRs) and a flash tank separator shown in Fig. 1.6. A feed
stream to the first CSTR F_{10} contains the reactant A which is converted into the
desired product B. The desired product B can then further react into an undesired

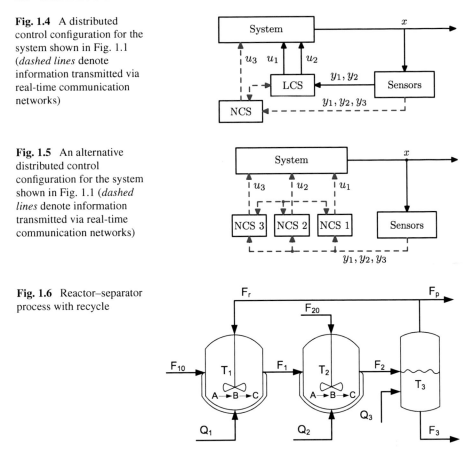

Fig. 1.4 A distributed control configuration for the system shown in Fig. 1.1 (*dashed lines* denote information transmitted via real-time communication networks)

Fig. 1.5 An alternative distributed control configuration for the system shown in Fig. 1.1 (*dashed lines* denote information transmitted via real-time communication networks)

Fig. 1.6 Reactor–separator process with recycle

side-product C. The effluent of the first CSTR along with additional fresh feed F_{20} makes up the inlet to the second CSTR. The reactions $A \to B$ and $B \to C$ take place in the two CSTRs in series before the effluent from CSTR 2 is fed to the flash tank. The overhead vapor from the flash tank is condensed and recycled to the first CSTR, and the bottom product stream is removed. A small portion of the overhead is purged before being recycled to the first CSTR.

The control objective is to stabilize the process at a desired operating steady-state and achieve an optimal level of closed-loop performance. To accomplish the control objective, we may design three local single loop controllers to manipulate the three heat inputs, Q_1, Q_2, Q_3, based on continuous temperature measurements of the three vessels. The three local controllers may be designed via proportional-integral-derivative (PID) control. This control configuration is shown in Fig. 1.7, which is the common traditional local control system configuration for a process shown in Fig. 1.6. This local control configuration corresponds to the control architecture shown in Fig. 1.1.

In the reactor–separator process, the additional information that we have access to because of additionally deployed networked sensors could be the species concen-

Fig. 1.7 Local control configuration for the reactor–separator process

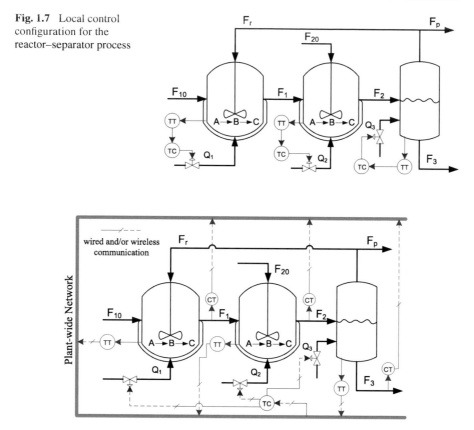

Fig. 1.8 A networked control configuration for the reactor–separator process. In this configuration, a networked control system is designed to replace the three local control loops in the local control configuration

tration measurements of each component in the three vessels. These measurements are subject to sampling delays and network transmission data package dropouts and they may not be available at every sampling time. To use the additional information, we may design an NCS to replace the three local control loops. This networked control configuration of the reactor–separator process is shown in Fig. 1.8 which corresponds to the control architecture shown in Fig. 1.2.

Instead of replacing the local control loops, an alternative to the above networked control configuration is to design an NCS to augment the local control loops to take advantage of the additional species concentration measurements as well as of the temperature measurements to adjust additional manipulated inputs, for instance, the feed flow rate to the second vessel, F_{20}. This networked control configuration of the reactor–separator process is shown in Fig. 1.9 which corresponds to the control architecture shown in Fig. 1.3.

Figure 1.10 shows a distributed control configuration for the reactor–separator process. In this design, two networked controllers are designed to manipulate

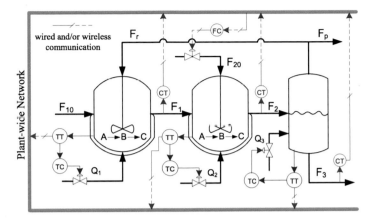

Fig. 1.9 A networked control configuration for the reactor–separator process. In this configuration, a networked control system in addition to the three local controllers is designed to improve the closed-loop performance

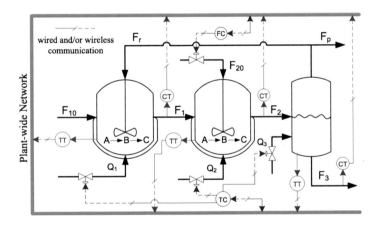

Fig. 1.10 A distributed control configuration for the reactor–separator process. In this configuration, the two networked controllers communicate via the plant-wide network to coordinate their actions

the three heat inputs and the feed flow rate to vessel 2, respectively, and communicate through the plant-wide network to exchange information and coordinate their actions. This control configuration corresponds to the one shown in Fig. 1.4.

Figure 1.11 shows the alternative distributed control configuration corresponding to Fig. 1.5 for the reactor–separator process. In this design, four networked controllers are designed to manipulate the four control inputs and communicate through the plant-wide network to exchange information and coordinate their actions.

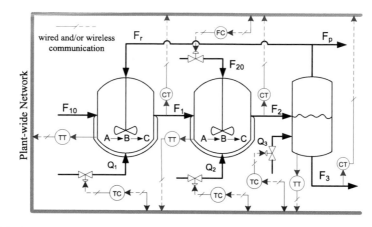

Fig. 1.11 A distributed control configuration for the reactor–separator process. In this configuration, four networked controllers are designed to manipulate the four control inputs and communicate via the plant-wide network to coordinate their actions

1.3 Background

Within control theory, the study of control over networks has attracted considerable attention in the literature (e.g., [7, 11, 69, 76, 103]) and early research focused on analyzing and scheduling real-time network traffic (e.g., [34, 96]). Research has also focused on the stability of network-based control systems. A common approach is to insert network behavior between the nodes of a conventional control loop, designed without taking the network behavior into account. More specifically, in [111], it was proposed to first design the controller using established techniques considering the network transparent, and then to analyze the effect of the network on closed-loop system stability and performance. This approach was further developed in [77] using a small gain analysis approach. In the last few years, however, several research papers have studied control using the IEEE 802.11 and Bluetooth wireless networks, see, for example, [85, 101, 115, 116] and the references therein. In the design and analysis of networked control systems, the most frequently studied problem considers control over a network having constant or time-varying delays. This network behavior is typical of communications over the Internet but does not necessarily represent the behavior of dedicated wireless networks in which the sensor, controller, and actuator nodes communicate directly with one another but might experience data losses. An appropriate framework to model lost data, is the use of asynchronous systems [29, 94, 99]. In this framework, data losses occur in an stochastic manner, and the process is considered to operate in an open-loop fashion when data is lost. The most destabilizing cause of packet loss is due to bursts of poor network performance in which case large groups of packets are lost nearly consecutively. A more detailed description of bursty network performance using a two-state Markov chain was considered in [81]. Modeling networks, using Markov chains results in describing the overall closed-loop system as a stochastic hybrid system [32]. Stability results have been presented for particular cases of stochastic hybrid systems (e.g., [29,

63]). However, these results do not directly address the problem of augmentation of dedicated, wired control systems with networked actuator and sensor devices to improve closed-loop performance.

With respect to other results on networked control, in [46], stability and disturbance attenuation issues for a class of linear networked control systems subject to data losses modeled as a discrete-time switched linear system with arbitrary switching was studied. In [35], (see also [3, 20, 28]), optimal control of linear time-invariant systems over unreliable communication links under different communication protocols (with and without acknowledgement of successful communication) was investigated and sufficient conditions for the existence of stabilizing control laws were derived. In [29], the stability properties of a class of networked control systems modeled as linear asynchronous systems was studied. Networked control systems in which the plant is modeled by a nonlinear system have received less attention. Limited access systems where each unit must compete with the others for access to the network have been studied in [77, 78, 110, 111] within a sampled-data system framework. In these works, practical stability of the system is guaranteed if the maximum time for which access to the network is not available is smaller than a given constant denoted as the maximum allowable transmission interval (MATI). A common theme of the above-mentioned works is that the controller is designed without taking into account the network dynamics and subsequently, the robustness of the closed-loop system in the presence of the network dynamics is studied. Furthermore, the importance of time delays in the context of networked control systems has also motivated significant research effort in modeling such delays and designing control systems to deal with them, primarily in the context of linear systems (e.g., [24, 45, 71, 112, 113, 118]).

In another recent line of work, Antsaklis and coworkers [70, 71] have proposed a strategy based on using an estimate of the state computed via the nominal model of the plant to decide the control input over the period of time in which feedback is lost between consecutively received measurements. In [70, 71], this framework was applied to optimize the bandwidth needed by a networked control system modeled as a sampled-data linear system with variable sampling rate. Other relevant works related to this approach include [74, 75], where the design of a linear output-feedback controller to stabilize a linear networked control system in the presence of delays, sampling and data losses was addressed. Within process control, important recent work on the subject of networked process control includes the development of a quasi-decentralized control framework for multi-unit plants that achieves the desired closed-loop objectives with minimal cross communication between the plant units [100]. In this work, the key idea is to embed in the local control system of each unit a set of dynamic models that provide an approximation of the interactions between a given unit and its neighbors in the plant when measurements are not transmitted through the plant-wide network and to update the state of each model using measurements from the corresponding unit when communication is reestablished. In addition to these works, fault diagnosis and fault-tolerant control methods that account for network-induced measurement errors have been developed in [26]. Finally, it is also important to note that within process control practice, wireless communication standards (e.g., ISA100 and WirelessHART) which are appropriate for chemical process industry applications have been developed based on the IEEE 802.15.4 standard [66] and applications of wireless field networks in the monitoring and control of chemical processes including heat exchangers and a phosphate fertilizer plant have been reported [119]. Despite these efforts, the problem of designing networked control systems that *explicitly account* for asynchronous and delayed measurements at both the design and implementation stages in the context of nonlinear systems, has received limited attention.

MPC has been widely used in the handling of measurement losses and delays because of its ability to predict the evolution of a system with time while accounting for the effect of data losses and delays. However, most of the available results deal with linear systems (e.g., [36, 49]). MPC is also a natural control framework to deal with the design of coordinated, distributed control systems because of its ability to handle input and state constraints, and also because it can account for the actions of other actuators in computing the control action of a given set of control actuators in real-time. With respect to available results in this direction, several distributed MPC (DMPC) methods have been proposed in the literature that deal with the coordination of separate MPCs that communicate in order to obtain optimal input trajectories in a distributed manner; see [8, 92, 95] for reviews of results in this area. More specifically, in [17], the problem of distributed control of dynamically coupled nonlinear systems that are subject to decoupled constraints was considered. In [37, 93], the effect of the coupling was modeled as a bounded disturbance compensated using a robust MPC formulation. In [98, 108], it was proven that through multiple communications between distributed controllers and using system-wide control objective functions, stability of the closed-loop system can be guaranteed for linear systems. In [39], DMPC of decoupled systems (a class of systems of relevance in the context of multi-agents systems) was studied. In [62], a DMPC algorithm was proposed under the main condition that the system is nonlinear, discrete-time and no information is exchanged between local controllers, and in [90], DMPC for nonlinear systems was studied from an input-to-state stability point of view. In [60, 61], a game theory based DMPC scheme for constrained linear systems was proposed. Previous work on MPC design for systems subject to asynchronous or delayed feedback has primarily focused on centralized MPC designs [27, 36, 49, 53, 72]. In addition to these works, control and monitoring of complex distributed systems with distributed intelligent agents were studied in [13, 84, 102]. Despite this progress, little attention has been given to the design of DMPC for systems subject to asynchronous or delayed measurements except in a recent work [22] where the issue of delays in the communication between distributed controllers was addressed.

1.4 Objectives and Organization of the Book

Motivated by the lack of general networked and distributed control methods for process systems, the broad objectives of this book are as follows:

1. To develop Lyapunov-based predictive control methods for nonlinear systems that provide an explicit characterization for the closed-loop stability region and account for the effect of asynchronous feedback and time-varying measurement delays.
2. To present a framework for the design of networked predictive control systems for nonlinear processes that naturally augment dedicated control systems with networked control systems.
3. To develop distributed predictive control methods for large-scale nonlinear process networks taking into account asynchronous measurements and time-varying delays as well as different sampling rates of measurements.
4. To illustrate the applications of the developed networked and distributed predictive control methods to nonlinear process networks and wind–solar energy generation systems.

The book is organized as follows. In Chap. 2, we first review some basic results on Lyapunov-based control, model predictive control and Lyapunov-based model predictive control (LMPC) of nonlinear systems and then present two Lyapunov-based model predictive control designs for systems subject to data losses and time-varying measurement delays. In order to guarantee the closed-loop stability, in the design of the LMPCs, constraints based on Lyapunov functions are incorporated. The theoretical results are illustrated through a chemical reactor example.

In Chap. 3, we present a two-tier networked control architecture to augment preexisting, point-to-point control systems with networked control systems, which take advantage of real-time wired or wireless sensor and actuator networks. Specifically, we will first present the two-tier networked control architecture for systems with continuous and asynchronous measurements; and then extend the results to include systems with continuous and asynchronous measurements which involve time-varying measurement delays. Two chemical process examples are used to illustrate the applicability and effectiveness of the two-tier control architecture. Moreover, the two-tier control architecture is also applied to the optimal management and operation of a standalone wind–solar energy generation system.

In Chap. 4, we focus on a class of distributed control problems that arise when new control systems which may use networked sensors and actuators are added to already operating control loops designed via MPC to improve closed-loop performance. To address this control problem, a distributed model predictive control method is introduced where the preexisting control system and the new control system are redesigned/designed via LMPC. The distributed control design stabilizes the closed-loop system, improves the closed-loop performance and allows handling input constraints. Furthermore, the distributed control design requires that these controllers communicate only once at each sampling time and is computationally more efficient compared to the corresponding centralized model predictive control design. The distributed control method is extended to include nonlinear systems subject to asynchronous and delayed measurements. The applicability and effectiveness of these distributed predictive control designs are illustrated through extensive simulations using a chemical plant example described by a nonlinear model.

In Chap. 5, we extend the results of Chap. 4 to distributed model predictive control of large-scale nonlinear systems in which several distinct sets of manipulated inputs are used to regulate the system. For each set of manipulated inputs, a different model predictive controller is used to compute the control actions, which is able to communicate with the rest of the controllers in making its decisions. We present two distributed control architectures designed via LMPC techniques. In the first architecture, the distributed controllers use a one-directional communication strategy, are evaluated in sequence and each controller is evaluated only once at each sampling time; in the second architecture, the distributed controllers utilize a bi-directional communication strategy, are evaluated in parallel and iterate to improve closed-loop performance. The case in which continuous state feedback is available to all the distributed controllers is first considered and then the results are extended to include large-scale nonlinear systems subject to asynchronous and delayed state feedback. The theoretical results are illustrated through a catalytic alkylation of benzene process example. Moreover, we also discuss how to handle disruptions in the communication between the distributed controllers by incorporating suitable feasibility problems for accepting/rejecting received information.

The designs of the distributed predictive control architectures in Chap. 5 are based on the assumptions that all the measurements of the system states are sampled simultaneously. In Chap. 6, we consider the design of a distributed predictive control system using multirate sampling for large-scale nonlinear uncertain systems composed of several coupled subsystems. Specifically, we assume that the states of each local subsystem can be divided into fast sampled states (which are available every sampling time) and slowly sampled states (which are available every several sampling times). The distributed predictive controllers are connected through a shared communication network and cooperate in an iterative fashion at time instants in which full system state measurements (both fast and slow) are available, to guarantee closed-loop stability. When local subsystem fast sampled state information is only available, the distributed controllers operate in a decentralized fashion to improve closed-loop performance. In the design of the distributed controllers, we also take into account bounded measurement noise, process disturbances and communication noise. The multirate distributed predictive control system is applied to a chemical reactor–separator process.

Chapter 7 summarizes the main results of the book and discusses future research directions in networked and distributed process control.

Chapter 2
Lyapunov-Based Model Predictive Control

2.1 Introduction

MPC, also known as receding horizon control (RHC), is a popular control strategy for the design of high performance model-based process control systems because of its ability to handle multi-variable interactions, constraints on control (manipulated) inputs and system states, and optimization requirements in a systematic manner. MPC is an online optimization-based approach, which takes advantage of a system model to predict its future evolution starting from the current system state along a given prediction horizon. Using model predictions, a future control input trajectory is optimized by minimizing a typically quadratic cost function involving penalties on the system states and control actions. To obtain finite dimensional optimization problems, MPC optimizes over a family of piecewise constant trajectories with a fixed sampling time and a finite prediction horizon. Once the optimization problem is solved, only the first manipulated input value is implemented and the rest of the trajectory is discarded; this optimization procedure is then repeated in the next sampling step [25, 91]. This is the so-called receding horizon scheme. The success of MPC in industrial applications (e.g., [25, 89]) has motivated numerous research investigations into the stability, robustness and optimality of model predictive controllers [65]. One important issue arising from these works is the difficulty in characterizing, a priori, the set of initial conditions starting from where controller feasibility and closed-loop stability are guaranteed. This issue motivated research on LMPC designs [67, 68] (see also [42, 86]) which allow for an explicit characterization of the stability region of the closed-loop system and lead to a reduced computational complexity of the controller optimization problem. Despite this progress, the adoption of communication networks in the control loops and the use of heterogeneous measurements motivate the development of MPC schemes that take data losses (or asynchronous feedback) and time-varying delays explicitly into account. However, little attention has been given to these issues except for a few results on MPC of linear systems with delays (e.g., [36, 49]).

Motivated by the above considerations, in this chapter, we adopt the LMPC framework [67, 68] and introduce modifications on the LMPC design both in the

P.D. Christofides et al., *Networked and Distributed Predictive Control*,
Advances in Industrial Control,
DOI 10.1007/978-0-85729-582-8_2, © Springer-Verlag London Limited 2011

optimization problem formulation and in the controller implementation to account for data losses and time-varying delays, respectively. The design of the LMPC is based on uniting receding horizon control with explicit Lyapunov-based nonlinear controller design techniques. In order to guarantee the closed-loop stability, in the design of the LMPCs, constraints based on Lyapunov functions are incorporated in the controller formulations. The theoretical results are illustrated through a chemical reactor example. The results of this chapter were first presented in [53, 72], and an application of the control methods to a continuous crystallizer can be found in [50].

2.2 Notation

Throughout this book, the operator $|\cdot|$ is used to denote the absolute value of a scalar and the operator $\|\cdot\|$ is used to denote Euclidean norm of a vector, while we use $\|\cdot\|_Q$ to denote the square of a weighted Euclidean norm, i.e., $\|x\|_Q = x^T Q x$ for all $x \in R^n$. A continuous function $\alpha : [0, a) \to [0, \infty)$ is said to belong to class $\mathcal{K}$ if it is strictly increasing and satisfies $\alpha(0) = 0$. A function $\beta(r, s)$ is said to be a class $\mathcal{K}\mathcal{L}$ function if, for each fixed s, $\beta(r, s)$ belongs to class $\mathcal{K}$ function with respect to r and, for each fixed r, $\beta(r, s)$ is decreasing with respect to s and $\beta(r, s) \to 0$ as $s \to 0$. The symbol Ω_r is used to denote the set $\Omega_r := \{x \in R^n : V(x) \le r\}$ where V is a scalar positive definite, continuous differentiable function and $V(0) = 0$, and the operator '/' denotes set subtraction, that is, $A/B := \{x \in R^n : x \in A, x \notin B\}$. The symbol $diag(v)$ denotes a square diagonal matrix whose diagonal elements are the elements of the vector v. The notation t_0 indicates the initial time instant. The set $\{t_{k\ge0}\}$ denotes a sequence of synchronous time instants such that $t_k = t_0 + k\Delta$ and $t_{k+i} = t_k + i\Delta$ where Δ is a fixed time interval and i is an integer. Similarly, the set $\{t_{a\ge0}\}$ denotes a sequence of asynchronous time instants such that the interval between two consecutive time instants is not fixed.

2.3 System Description

Consider nonlinear systems described by the following state-space model:

$$\dot{x}(t) = f\big(x(t), u(t), w(t)\big), \tag{2.1}$$

where $x(t) \in R^n$ denotes the vector of state variables, $u(t) \in R^m$ denotes the vector of control (manipulated) input variables, $w(t) \in R^w$ denotes the vector of disturbance variables and f is a locally Lipschitz vector function on $R^n \times R^m \times R^w$ such that $f(0, 0, 0) = 0$. This implies that the origin is an equilibrium point for the nominal system (i.e., system of Eq. 2.1 with $w(t) \equiv 0$ for all t) with $u = 0$.

The input vector is restricted to be in a nonempty convex set $U \subseteq R^m$ which is defined as follows:

$$U := \big\{u \in R^m : \|u\| \le u^{\max}\big\}, \tag{2.2}$$

where $u^{\max}$ is the magnitude of the input constraint.

The disturbance vector is bounded, that is, $w(t) \in W$ where:

$$W := \left\{ w \in R^w : \|w\| \le \theta, \theta > 0 \right\} \tag{2.3}$$

with θ being a known positive real number. The vector of uncertain variables, $w(t)$, is introduced into the model in order to account for the occurrence of uncertainty in the values of the process parameters and the influence of disturbances in process control applications.

Remark 2.1 Note that the assumption that f is a locally Lipschitz vector function is a reasonable assumption for most of chemical process models.

2.4 Lyapunov-Based Control

We assume that there exists a feedback control law $u(t) = h(x(t))$ which satisfies the input constraint on u for all x inside a given stability region and renders the origin of the nominal closed-loop system asymptotically stable. This assumption is essentially equivalent to the assumption that the nominal system is stabilizable or that there exists a Lyapunov function for the nominal system or that the pair (A, B) in the case of linear systems is stabilizable. Using converse Lyapunov theorems [11, 40, 48, 64], this assumption implies that there exist functions $\alpha_i(\cdot)$, $i = 1, 2, 3, 4$ of class $\mathcal{K}$ and a continuously differentiable Lyapunov function $V(x)$ for the nominal closed-loop system, that satisfy the following inequalities:

$$\alpha_1(\|x\|) \le V(x) \le \alpha_2(\|x\|), \tag{2.4}$$

$$\frac{\partial V(x)}{\partial x} f(x, h(x), 0) \le -\alpha_3(\|x\|), \tag{2.5}$$

$$\left\| \frac{\partial V(x)}{\partial x} \right\| \le \alpha_4(\|x\|), \tag{2.6}$$

$$h(x) \in U \tag{2.7}$$

for all $x \in O \subseteq R^n$ where O is an open neighborhood of the origin. We denote the region $\Omega_\rho \subseteq O$ as the stability region of the closed-loop system under the control $u = h(x)$. Note that explicit stabilizing control laws that provide explicitly defined regions of attraction for the closed-loop system have been developed using Lyapunov techniques for specific classes of nonlinear systems, particularly input-affine nonlinear systems; the reader may refer to [2, 11, 41, 97] for results in this area including results on the design of bounded Lyapunov-based controllers by taking explicitly into account constraints for broad classes of nonlinear systems [18, 19, 47].

By continuity, the local Lipschitz property assumed for the vector field $f(x, u, w)$, the fact that the manipulated input u is bounded in a convex set and the continuous differentiable property of the Lyapunov function V, there exists positive constants M, L_w, L_x and L_x' such that:

$$\|f(x, u, w)\| \leq M, \tag{2.8}$$

$$\|f(x, u, w) - f(x', u, 0)\| \leq L_w \|w\| + L_x \|x - x'\|, \tag{2.9}$$

$$\left\|\frac{\partial V(x)}{\partial x} f(x, u, 0) - \frac{\partial V(x')}{\partial x} f(x', u, 0)\right\| \leq L_x' \|x - x'\| \tag{2.10}$$

for all $x, x' \in \Omega_\rho$, $u \in U$ and $w \in W$. These constants will be used in characterizing the stability properties of the system of Eq. 2.1 under LMPC designs.

Remark 2.2 Note that while there are currently no general methods for constructing Lyapunov functions for general nonlinear systems, for broad classes of nonlinear models arising in the context of chemical process control applications, quadratic Lyapunov functions are widely used and provide very good estimates of closed-loop stability regions.

Remark 2.3 Note that the inequalities of Eqs. 2.4–2.10 are derived from the basic assumptions (i.e., Lipschitz vector field and existence of a stabilizing Lyapunov-based controller). The various constants involved in the upper bounds are not assumed to be arbitrarily small.

2.5 Model Predictive Control

MPC is widely adopted in industry as an effective approach to deal with large multivariable constrained control problems. The main idea of MPC is to choose control actions by repeatedly solving an online constrained optimization problem, which aims at minimizing a performance index over a finite prediction horizon based on predictions obtained by a system model. In general, an MPC design is composed of three components:

1. A model of the system. This model is used to predict the future evolution of the system in open-loop and the efficiency of the calculated control actions of an MPC depends highly on the accuracy of the model.
2. A performance index over a finite horizon. This index will be minimized subject to constraints imposed by the system model, restrictions on control inputs and system state and other considerations at each sampling time to obtain a trajectory of future control inputs.
3. A receding horizon scheme. This scheme introduces the notion of feedback into the control law to compensate for disturbances and modeling errors.

Consider the control of the system of Eq. 2.1 and assume that the state measurements of the system of Eq. 2.1 are available at synchronous sampling time instants

$\{t_{k\geq 0}\}$, a standard MPC is formulated as follows [25]:

$$\min_{u \in S(\varDelta)} \int_{t_k}^{t_{k+N}} \left[\left\| \tilde{x}(\tau) \right\|_{Q_c} + \left\| u(\tau) \right\|_{R_c} \right] d\tau + F\big(x(t_{k+N})\big), \qquad (2.11)$$

$$\text{s.t.} \quad \dot{\tilde{x}}(t) = f\big(\tilde{x}(t), u(t), 0\big), \qquad (2.12)$$

$$u(t) \in U, \qquad (2.13)$$

$$\tilde{x}(t_k) = x(t_k), \qquad (2.14)$$

where $S(\varDelta)$ is the family of piece-wise constant functions with sampling period $\varDelta$, N is the prediction horizon, Q_c and R_c are strictly positive definite symmetric weighting matrices, $\tilde{x}$ is the predicted trajectory of the nominal system due to control input u with initial state $x(t_k)$ at time t_k, and $F(\cdot)$ denotes the terminal penalty.

The optimal solution to the MPC optimization problem defined by Eqs. 2.11–2.14 is denoted as $u^*(t|t_k)$ which is defined for $t \in [t_k, t_{k+N})$. The first step value of $u^*(t|t_k)$ is applied to the closed-loop system for $t \in [t_k, t_{k+1})$. At the next sampling time t_{k+1}, when a new measurement of the system state $x(t_{k+1})$ is available, the control evaluation and implementation procedure is repeated. The manipulated input of the system of Eq. 2.1 under the control of the MPC of Eqs. 2.11–2.14 is defined as follows:

$$u(t) = u^*(t|t_k), \quad \forall t \in [t_k, t_{k+1}), \qquad (2.15)$$

which is the standard receding horizon scheme.

In the MPC formulation of Eqs. 2.11–2.14, Eq. 2.11 defines a performance index or cost index that should be minimized. In addition to penalties on the state and control actions, the index may also include penalties on other considerations; for example, the rate of change of the inputs. Equation 2.12 is the nominal model of the system of Eq. 2.1 which is used in the MPC to predict the future evolution of the system. Equation 2.13 takes into account the constraint on the control input, and Eq. 2.14 provides the initial state for the MPC which is a measurement of the actual system state. Note that in the above MPC formulation, state constraints are not considered but can be readily taken into account.

It is well known that the MPC of Eqs. 2.11–2.14 is not necessarily stabilizing. To achieve closed-loop stability, different approaches have been proposed in the literature. One class of approaches is to use infinite prediction horizons or well-designed terminal penalty terms; please see [6, 65] for surveys of these approaches. Another class of approaches is to impose stability constraints in the MPC optimization problem [1, 4, 65]. There are also efforts focusing on getting explicit stabilizing MPC laws using offline computations [59]. However, the implicit nature of MPC control law makes it very difficult to explicitly characterize, a priori, the admissible initial conditions starting from where the MPC is guaranteed to be feasible and stabilizing. In practice, the initial conditions are usually chosen in an ad hoc fashion and tested through extensive closed-loop simulations.

2.6 Lyapunov-Based Model Predictive Control

In this section, we introduce the LMPC design proposed in [67, 68] which allows for an explicit characterization of the stability region and guarantees controller feasibility and closed-loop stability.

For the predictive control of the system of Eq. 2.1, the LMPC is designed based on an existing explicit control law $h(x)$ which is able to stabilize the closed-loop system and satisfies the conditions of Eqs. 2.4–2.7. The formulation of the LMPC is as follows:

$$\min_{u \in S(\Delta)} \int_{t_k}^{t_{k+N}} \left[\left\| \tilde{x}(\tau) \right\|_{Q_c} + \left\| u(\tau) \right\|_{R_c} \right] d\tau, \tag{2.16}$$

$$\text{s.t.}\quad \dot{\tilde{x}}(t) = f\big(\tilde{x}(t), u(t), 0\big), \tag{2.17}$$

$$u(t) \in U, \tag{2.18}$$

$$\tilde{x}(t_k) = x(t_k), \tag{2.19}$$

$$\frac{\partial V(x(t_k))}{\partial x} f\big(x(t_k), u(t_k), 0\big) \leq \frac{\partial V(x(t_k))}{\partial x} f\big(x(t_k), h(x(t_k)), 0\big), \tag{2.20}$$

where $V(x)$ is a Lyapunov function associated with the nonlinear control law $h(x)$. The optimal solution to this LMPC optimization problem is denoted as $u_l^*(t|t_k)$ which is defined for $t \in [t_k, t_{k+N})$. The manipulated input of the system of Eq. 2.1 under the control of the LMPC of Eqs. 2.16–2.20 is defined as follows:

$$u(t) = u_l^*(t|t_k), \quad \forall t \in [t_k, t_{k+1}), \tag{2.21}$$

which implies that this LMPC also adopts a standard receding horizon strategy.

In the LMPC defined by Eqs. 2.16–2.20, the constraint of Eq. 2.20 guarantees that the value of the time derivative of the Lyapunov function, $V(x)$, at time t_k is smaller than or equal to the value obtained if the nonlinear control law $u = h(x)$ is implemented in the closed-loop system in a sample-and-hold fashion. This is a constraint that allows one to prove (when state measurements are available every synchronous sampling time) that the LMPC inherits the stability and robustness properties of the nonlinear control law $h(x)$ when it is applied in a sample-and-hold fashion.

One of the main properties of the LMPC of Eqs. 2.16–2.20 is that it possesses the same stability region Ω_ρ as the nonlinear control law $h(x)$, which implies that the origin of the closed-loop system is guaranteed to be stable and the LMPC is guaranteed to be feasible for any initial state inside Ω_ρ when the sampling time Δ and the disturbance upper bound θ are sufficiently small. Note that the region Ω_ρ can be explicitly characterized; please refer to Sect. 2.4 for more discussion on this issue. The stability property of the LMPC is inherited from the nonlinear control law $h(x)$ when it is applied in a sample-and-hold fashion; please see [14, 79] for results on sampled-data systems. The feasibility property of the LMPC is also guaranteed by the nonlinear control law $h(x)$ since $u = h(x)$ is a feasible solution to the optimization problem of Eqs. 2.16–2.20. The main advantage of the LMPC approach with respect to the nonlinear control law $h(x)$ is that optimality considerations can be taken explicitly into account (as well as constraints on the inputs and the states [68]) in the computation of the control actions within an online optimization framework while improving the closed-loop performance of the system.

Remark 2.4 Since the closed-loop stability and feasibility of the LMPC of Eqs. 2.16–2.20 are guaranteed by the nonlinear control law $h(x)$, it is unnecessary to use a terminal penalty term in the cost index (see Eq. 2.16 and compare it with Eq. 2.11) and the length of the horizon N does not affect the stability of the closed-loop system but it affects the closed-loop performance.

2.7 LMPC with Asynchronous Feedback

In this section, we modify the LMPC introduced in the previous section to take into account data losses or asynchronous measurements, both in the optimization problem formulation and in the controller implementation. In this LMPC scheme, when feedback is lost, instead of setting the control actuator outputs to zero or to the last available values, the actuators implement the last optimal input trajectory evaluated by the controller (this requires that the actuators must store in memory the last optimal input trajectory received). The LMPC is designed based on a nonlinear control law which is able to stabilize the closed-loop system and inherits the stability and robustness properties in the presence of uncertainty and data losses of the nonlinear controller, while taking into account optimality considerations. Specifically, the LMPC scheme allows for an explicit characterization of the stability region, guarantees practical stability in the absence of data losses or asynchronous measurements, and guarantees that the stability region is an invariant set for the closed-loop system under data losses or asynchronous measurements if the maximum time in which the loop is open is shorter than a given constant that depends on the parameters of the system and the nonlinear control law that is used to formulate the optimization problem. A schematic diagram of the considered closed-loop system is shown in Fig. 2.1.

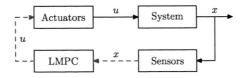

Fig. 2.1 LMPC design for systems subject to data losses. *Solid lines* denote point-to-point, wired communication links; *dashed lines* denote networked communication and/or asynchronous sampling/actuation

2.7.1 Modeling of Data Losses/Asynchronous Measurements

We assume that feedback of the state of the system of Eq. 2.1, $x(t)$, is available at asynchronous time instants t_a where $\{t_{a\geq0}\}$ is a random increasing sequence of times; that is, the intervals between two consecutive instants are not fixed. The distribution of $\{t_{a\geq0}\}$ characterizes the time the feedback loop is closed or the time needed to obtain a new state measurement. In general, if there exists the possibility of arbitrarily large periods of time in which feedback is not available, then it is not possible to provide guaranteed stability properties, because there exists a nonzero probability that the system operates in open-loop for a period of time large enough for the state to leave the stability region. In order to study the stability properties in a deterministic framework, we assume that there exists an upper bound T_m on the interval between two successive time instants in which the feedback loop is closed or new state measurements are available, that is:

$$\max_{a}\{t_{a+1} - t_a\} \leq T_m. \tag{2.22}$$

This assumption is reasonable from process control and networked control systems perspectives [69, 78, 110, 111] and allows us to study deterministic notions of stability. This model of feedback/measurements is of relevance to systems subject to asynchronous measurement samplings and to networked control systems, where the asynchronous property is introduced by data losses in the communication network connecting the sensors/actuators and the controllers.

2.7.2 LMPC Formulation with Asynchronous Feedback

When feedback is lost, most approaches set the control input to zero or to the last implemented value. Instead, in this LMPC for systems subject to data losses, when feedback is lost, we take advantage of the MPC scheme to update the input based on a prediction obtained using the system model. This is achieved using the following implementation strategy:

1. At a sampling time, t_a, when the feedback loop is closed (i.e., the current system state $x(t_a)$ is available for the controller and the controller can send information to the actuators), the LMPC evaluates the optimal future input trajectory $u(t)$ for $t \in [t_a, t_a + N\Delta)$.
2. The LMPC sends the entire optimal input trajectory (i.e., $u(t)\ \forall t \in [t_a, t_a + N\Delta)$) to the actuators.
3. The actuators implement the input trajectory until the feedback loop is closed again at the next sampling time t_{a+1}; that is, the actuators implement $u(t)$ in $t \in [t_a, t_{a+1})$.
4. When a new measurement is received ($a \leftarrow a + 1$), go to Step 1.

In this implementation strategy, when the state is not available, or the data sent from the controller to the actuators is lost, the actuators keep implementing the last received optimal trajectory. If data is lost for a period larger than the prediction horizon, the actuators set the inputs to the last implemented values or to fixed values. This strategy is a receding horizon scheme, which takes into account that data losses may occur. This strategy is motivated by the fact that when no feedback is available, a reasonable estimate of the future evolution of the system is given by the nominal trajectory. The LMPC design taking into account data losses/asynchronous measurements, therefore modifies the standard implementation scheme of switching off the actuators ($u = 0$) or setting the actuators to nominal values or to the last computed input values. The idea of using the model to predict the evolution of the system when no feedback is possible has also been used in the context of sampled-data linear systems, see [70, 71, 74, 75]. The actuators not only receive and implement given inputs, but must also be able to store future trajectories to implement them in case data losses occur. This means that to handle data losses, not only the control algorithms must be modified, but also the control actuator hardware that implements the control actions.

When data losses are present in the feedback loop, the existing LMPC schemes [42, 67, 68, 86] can not guarantee the closed-loop stability no matter whether the actuators keep the inputs at the last values, set the inputs to constant values, or keep on implementing the previously evaluated input trajectories. In particular, there is no guarantee that the LMPC optimization problems will be feasible for all time, i.e., that the state will remain inside the stability region for all time. In the LMPC design of Eqs. 2.16–2.20, the constraint of Eq. 2.20 only takes into account the first prediction step and does not restrict the behavior of the system after the first step. If no additional constraints are included in the optimization problem, no claims on the closed-loop behavior of the system can be made. For this reason, when data losses are taken into account, the constraints of the LMPC problem have to be modified. The LMPC that takes into account data losses in an explicit way is based on the following finite horizon constrained optimal control problem:

$$\min_{u \in S(\Delta)} \int_{t_a}^{t_a + N\Delta} \left[\left\| \tilde{x}(\tau) \right\|_{Q_c} + \left\| u(\tau) \right\|_{R_c} \right] d\tau, \tag{2.23}$$

$$\text{s.t.} \quad \dot{\tilde{x}}(t) = f\big(\tilde{x}(t), u(t), 0\big), \tag{2.24}$$

$$\dot{\hat{x}}(t) = f\big(\hat{x}(t), h\big(\hat{x}(t_a + j\Delta)\big), 0\big), \quad \forall t \in \big[t_a + j\Delta, t_a + (j+1)\Delta\big), \tag{2.25}$$

$$u(t) \in U, \tag{2.26}$$

$$\tilde{x}(t_a) = \hat{x}(t_a) = x(t_a), \tag{2.27}$$

$$V\big(\tilde{x}(t)\big) \leq V\big(\hat{x}(t)\big), \quad \forall t \in [t_a, t_a + N_R\Delta), \tag{2.28}$$

where $\hat{x}(t)$ is the trajectory of the nominal system under the nonlinear control law $u = h(\hat{x}(t))$ when it is implemented in a sample-and-hold fashion, $j = 0, 1, \ldots, N - 1$, and N_R is the smallest integer satisfying $N_R\Delta \geq T_m$. This optimization problem does not depend on the uncertainty and assures that the LMPC inherits the properties of the nonlinear control law $h(x)$. To take full advantage of the use of the nominal model in the computation of the control action, the prediction horizon should be chosen in a way such that $N \geq N_R$.

The optimal solution to the LMPC optimization problem of Eqs. 2.23–2.28 is denoted as $u_a^*(t|t_a)$ which is defined for $t \in [t_a, t_a + N\Delta)$. The manipulated input of the system of Eq. 2.1 under the LMPC of Eqs. 2.23–2.28 is defined as follows:

$$u(t) = u_a^*(t|t_a), \quad \forall t \in [t_a, t_{a+1}), \tag{2.29}$$

where t_{a+1} is the next time instant in which the feedback loop will be closed again. This is a modified receding horizon scheme which takes advantage of the predicted input trajectory in the case of data losses.

In the design of the LMPC of Eqs. 2.23–2.28, the constraint of Eq. 2.25 is used to generate a system state trajectory under the nonlinear control law $u = h(x)$ implemented in a sample-and-hold fashion; this trajectory is used as a reference trajectory to construct the Lyapunov-based constraint of Eq. 2.28 which is required to be satisfied for a time period which covers the maximum possible open-loop operation time T_m. This Lyapunov-based constraint allows one to prove the closed-loop stability in the presence of data losses in the closed-loop system.

Remark 2.5 The LMPC of Eqs. 2.23–2.28 optimizes a cost function, subject to a set of constraints defined by the state trajectory corresponding to the nominal system in closed-loop. This allows us to formulate an LMPC problem that does not depend on the uncertainty and so it is of manageable computational complexity.

2.7.3 Stability Properties

The LMPC of Eqs. 2.23–2.28 computes the control input u applied to the system of Eq. 2.1 in a way such that in the closed-loop system, the value of the Lyapunov function at time instant t_a (i.e., $V(x(t_a))$) is a decreasing sequence of values with a lower bound. Following Lyapunov arguments, this property guarantees practical stability of the closed-loop system. This is achieved due to the constraint of Eq. 2.28. This property is summarized in Theorem 2.1 below. To state this theorem, we need the following propositions.

Proposition 2.1 *Consider the nominal sampled trajectory $\hat{x}(t)$ of the system of Eq. 2.1 in closed-loop for a controller $h(x)$, which satisfies the conditions of Eqs. 2.4–2.7, obtained by solving recursively:*

$$\dot{\hat{x}}(t) = f\big(\hat{x}(t), h\big(\hat{x}(t_k)\big), 0\big), \quad t \in [t_k, t_{k+1}), \tag{2.30}$$

where $t_k = t_0 + k\Delta, k = 0, 1, \ldots$ Let $\Delta, \varepsilon_s > 0$ and $\rho > \rho_s > 0$ satisfy:

$$-\alpha_3\big(\alpha_2^{-1}(\rho_s)\big) + L_x' M \Delta \leq -\varepsilon_s/\Delta. \tag{2.31}$$

Then if $\rho_{\min} < \rho$ where:

$$\rho_{\min} = \max\left\{V\left(\hat{x}(t+\Delta)\right) : V\left(\hat{x}(t)\right) \leq \rho_s\right\} \tag{2.32}$$

and $\hat{x}(t_0) \in \Omega_\rho$, the following inequality holds:

$$V\left(\hat{x}(t)\right) \leq V\left(\hat{x}(t_k)\right), \quad \forall t \in [t_k, t_{k+1}), \tag{2.33}$$

$$V\left(\hat{x}(t_k)\right) < \max\left\{V\left(\hat{x}(t_0)\right) - k\varepsilon_s, \rho_{\min}\right\}. \tag{2.34}$$

Proof Following the definition of $\hat{x}(t)$, the time derivative of the Lyapunov function $V(x)$ along the trajectory $\hat{x}(t)$ of the system of Eq. 2.1 in $t \in [t_k, t_{k+1})$ is given by:

$$\dot{V}\left(\hat{x}(t)\right) = \frac{\partial V(\hat{x}(t))}{\partial x} f\left(\hat{x}(t), h\left(\hat{x}(t_k)\right), 0\right). \tag{2.35}$$

Adding and subtracting $\frac{\partial V(\hat{x}(t_k))}{\partial x} f(\hat{x}(t_k), h(\hat{x}(t_k)), 0)$ and taking into account Eq. 2.5, we obtain:

$$\dot{V}\left(\hat{x}(t)\right) \leq -\alpha_3\left(\|\hat{x}(t_k)\|\right) + \frac{\partial V(\hat{x}(t))}{\partial x} f\left(\hat{x}(t), h\left(\hat{x}(t_k)\right), 0\right)$$

$$- \frac{\partial V(\hat{x}(t_k))}{\partial x} f\left(\hat{x}(t_k), h\left(\hat{x}(t_k)\right), 0\right). \tag{2.36}$$

From the Lipschitz property of Eq. 2.10 and the above inequality of Eq. 2.36, we have that:

$$\dot{V}\left(\hat{x}(t)\right) \leq -\alpha_3\left(\alpha_2^{-1}(\rho_s)\right) + L_x'\|\hat{x}(t) - \hat{x}(t_k)\| \tag{2.37}$$

for all $\hat{x}(t_k) \in \Omega_\rho/\Omega_{\rho_s}$. Taking into account the Lipschitz property of Eq. 2.8 and the continuity of $\hat{x}(t)$, the following bound can be written for all $t \in [t_k, t_{k+1})$:

$$\|\hat{x}(t) - \hat{x}(t_k)\| \leq M\Delta. \tag{2.38}$$

Using the expression of Eq. 2.38, we obtain the following bound on the time derivative of the Lyapunov function for $t \in [t_k, t_{k+1})$, for all initial states $\hat{x}(t_k) \in \Omega_\rho/\Omega_{\rho_s}$:

$$\dot{V}\left(\hat{x}(t)\right) \leq -\alpha_3\left(\alpha_2^{-1}(\rho_s)\right) + L_x'M\Delta. \tag{2.39}$$

If the condition of Eq. 2.31 is satisfied, then $\dot{V}(\hat{x}(t)) \leq -\varepsilon_s/\Delta$. Integrating this bound on $t \in [t_k, t_{k+1})$ we obtain that the inequality of Eq. 2.33 holds. Using Eq. 2.33 recursively, it is proved that, if $x(t_0) \in \Omega_\rho/\Omega_{\rho_s}$, the state converges to Ω_{ρ_s} in a finite number of sampling times without leaving the stability region. Once the state converges to $\Omega_{\rho_s} \subseteq \Omega_{\rho_{\min}}$, it remains inside $\Omega_{\rho_{\min}}$ for all times. This statement holds because of the definition of $\rho_{\min}$ in Eq. 2.32. $\square$

Proposition 2.1 ensures that if the nominal system under the control $u = h(x)$ implemented in a sample-and-hold fashion with state feedback every sampling time starts in the region Ω_ρ, then it is ultimately bounded in $\Omega_{\rho_{\min}}$. The following Proposition 2.2 provides an upper bound on the deviation of the system state trajectory obtained using the nominal model of Eq. 2.1, from the closed-loop state trajectory of the system of Eq. 2.1 under uncertainty (i.e., $w(t) \neq 0$) when the same control actions are applied.

Proposition 2.2 *Consider the systems*:

$$\dot{x}_a(t) = f\big(x_a(t), u(t), w(t)\big), \tag{2.40}$$

$$\dot{x}_b(t) = f\big(x_b(t), u(t), 0\big) \tag{2.41}$$

with initial states $x_a(t_0) = x_b(t_0) \in \Omega_\rho$. *There exists a class* $\mathcal{K}$ *function* $f_W(\cdot)$ *such that*:

$$\big\| x_a(t) - x_b(t) \big\| \leq f_W(t - t_0), \tag{2.42}$$

for all $x_a(t), x_b(t) \in \Omega_\rho$ *and all* $w(t) \in W$ *with*:

$$f_W(\tau) = \frac{L_w\theta}{L_x}\big(e^{L_x\tau} - 1\big). \tag{2.43}$$

Proof Define the error vector as $e(t) = x_a(t) - x_b(t)$. The time derivative of the error is given by:

$$\dot{e}(t) = f\big(x_a(t), u(t), w(t)\big) - f\big(x_b(t), u(t), 0\big). \tag{2.44}$$

From the Lipschitz property of Eq. 2.9, the following inequality holds:

$$\big\|\dot{e}(t)\big\| \leq L_w\big\|w(t)\big\| + L_x\big\|x_a(t) - x_b(t)\big\| \leq L_w\theta + L_x\big\|e(t)\big\| \tag{2.45}$$

for all $x_a(t), x_b(t) \in \Omega_\rho$ and $w(t) \in W$. Integrating $\|\dot{e}(t)\|$ with initial condition $e(t_0) = 0$ (recall that $x_a(t_0) = x_b(t_0)$), the following bound on the norm of the error vector is obtained:

$$\big\|e(t)\big\| \leq \frac{L_w\theta}{L_x}\big(e^{L_x(t-t_0)} - 1\big). \tag{2.46}$$

This implies that the inequality of Eq. 2.42 holds for:

$$f_W(\tau) = \frac{L_w\theta}{L_x}\big(e^{L_x\tau} - 1\big) \tag{2.47}$$

which proves this proposition.   □

Proposition 2.3 below bounds the difference between the magnitudes of the Lyapunov function of two states in Ω_ρ.

Proposition 2.3 *Consider the Lyapunov function $V(\cdot)$ of the system of Eq. 2.1. There exists a quadratic function $f_V(\cdot)$ such that:*

$$V(x) \leq V(x') + f_V(\|x - x'\|) \tag{2.48}$$

for all $x, x' \in \Omega_\rho$ where:

$$f_V(s) = \alpha_4(\alpha_1^{-1}(\rho))s + M_v s^2 \tag{2.49}$$

with $M_v > 0$.

Proof Since the Lyapunov function $V(x)$ is continuous and bounded on compact sets, there exists a positive constant M_v such that a Taylor series expansion of V around x' yields:

$$V(x) \leq V(x') + \frac{\partial V(x')}{\partial x}\|x - x'\| + M_v\|x - x'\|^2, \quad \forall x, x' \in \Omega_\rho. \tag{2.50}$$

Note that the term $M_v\|x - x'\|^2$ bounds the high order terms of the Taylor series of $V(x)$ for $x, x' \in \Omega_\rho$. Taking into account Eq. 2.6, the following bound for $V(x)$ is obtained:

$$V(x) \leq V(x') + \alpha_4(\alpha_1^{-1}(\rho))\|x - x'\| + M_v\|x - x'\|^2, \quad \forall x, x' \in \Omega_\rho, \tag{2.51}$$

which proves this proposition. ☐

In Theorem 2.1 below, we provide sufficient conditions under which the LMPC design of Eqs. 2.23–2.28 guarantees that the state of the closed-loop system of Eq. 2.1 is ultimately bounded in a region that contains the origin.

Theorem 2.1 *Consider the system of Eq. 2.1 in closed-loop, with the loop closing at asynchronous time instants $\{t_{a \geq 0}\}$ that satisfy the condition of Eq. 2.22, under the LMPC of Eqs. 2.23–2.28 based on a controller $h(x)$ that satisfies the conditions of Eqs. 2.4–2.7. Let $\Delta, \varepsilon_s > 0$, $\rho > \rho_{min} > 0$, $\rho > \rho_s > 0$ and $N \geq N_R \geq 1$ satisfy the condition of Eq. 2.31 and the following inequality:*

$$-N_R \varepsilon_s + f_V(f_W(N_R \Delta)) < 0 \tag{2.52}$$

with $f_V(\cdot)$ and $f_W(\cdot)$ defined in Eqs. 2.49 and 2.43, respectively, and N_R being the smallest integer satisfying $N_R \Delta \geq T_m$. If $x(t_0) \in \Omega_\rho$, then $x(t)$ is ultimately bounded in $\Omega_{\rho_a} \subseteq \Omega_\rho$ where:

$$\rho_a = \rho_{min} + f_V(f_W(N_R \Delta)) \tag{2.53}$$

with ρ_{min} defined as in Eq. 2.32.

Proof In order to prove that the closed-loop system is ultimately bounded in a region that contains the origin, we prove that $V(x(t_a))$ is a decreasing sequence of values with a lower bound. The proof is divided into two parts.

Part 1: In this part, we prove that the stability results stated in Theorem 2.1 hold in the case that $t_{a+1} - t_a = T_m$ for all a and $T_m = N_R \Delta$. This case corresponds to the worst possible situation in the sense that the LMPC needs to operate in open-loop for the maximum possible amount of time. In order to simplify the notation, we assume that all the notations used in this proof refer to the final solution of the LMPC of Eqs. 2.23–2.28 solved at time t_a. By Proposition 2.1 and the fact that $t_{a+1} = t_a + N_R \Delta$, the following inequality can be obtained:

$$V\big(\hat{x}(t_{a+1})\big) \leq \max\big\{V\big(\hat{x}(t_a)\big) - N_R \varepsilon_s, \rho_{\min}\big\}. \tag{2.54}$$

From the constraint of Eq. 2.28, the inequality of Eq. 2.54 and taking into account the fact that $\hat{x}(t_a) = \tilde{x}(t_a) = x(t_a)$, the following inequality can be written:

$$V\big(\tilde{x}(t_{a+1})\big) \leq \max\big\{V\big(x(t_a)\big) - N_R \varepsilon_s, \rho_{\min}\big\}. \tag{2.55}$$

When $x(t) \in \Omega_\rho$ for all times (this point will be proved below), we can apply Proposition 2.3 to obtain the following inequality:

$$V\big(x(t_{a+1})\big) \leq V\big(\tilde{x}(t_{a+1})\big) + f_V\big(\|\tilde{x}(t_{a+1}) - x(t_{a+1})\|\big). \tag{2.56}$$

Applying Proposition 2.2, we obtain the following upper bound on the deviation of $\tilde{x}(t)$ from $x(t)$:

$$\|x(t_{a+1}) - \tilde{x}(t_{a+1})\| \leq f_W(N_R \Delta). \tag{2.57}$$

From the inequalities of Eqs. 2.56 and 2.57, the following upper bound on $V(x(t_{a+1}))$ can be written:

$$V\big(x(t_{a+1})\big) \leq V\big(\tilde{x}(t_{a+1})\big) + f_V\big(f_W(N_R \Delta)\big). \tag{2.58}$$

Using the inequality of Eq. 2.55, we can rewrite the inequality of Eq. 2.58 as follows:

$$V\big(x(t_{a+1})\big) \leq \max\big\{V\big(x(t_a)\big) - N_R \varepsilon_s, \rho_{\min}\big\} + f_V\big(f_W(N_R \Delta)\big). \tag{2.59}$$

If the condition of Eq. 2.52 is satisfied, from the inequality of Eq. 2.59, we know that there exists $\varepsilon_w > 0$ such that the following inequality holds:

$$V\big(x(t_{a+1})\big) \leq \max\big\{V\big(x(t_a)\big) - \varepsilon_w, \rho_a\big\}, \tag{2.60}$$

which implies that if $x(t_a) \in \Omega_\rho / \Omega_{\rho_a}$, then $V(x(t_{a+1})) < V(x(t_a))$, and if $x(t_a) \in \Omega_{\rho_a}$, then $V(x(t_{a+1})) \leq \rho_a$.

Because $f_W(\cdot)$ and $f_V(\cdot)$ are strictly increasing functions of their arguments and $f_V(\cdot)$ is convex (see Propositions 2.2 and 2.3 for the expressions of $f_W(\cdot)$ and $f_V(\cdot)$), the inequality of Eq. 2.60 also implies that:

$$V\big(x(t)\big) \leq \max\big\{V\big(x(t_a)\big), \rho_a\big\}, \quad \forall t \in [t_a, t_{a+1}). \tag{2.61}$$

Using the inequality of Eq. 2.61 recursively, it can be proved that if $x(t_0) \in \Omega_\rho$, then the closed-loop trajectories of the system of Eq. 2.1 under the LMPC of Eqs. 2.23–2.28 stay in Ω_ρ for all times (i.e., $x(t) \in \Omega_\rho$, $\forall t$). Moreover, it can be proved that if $x(t_0) \in \Omega_\rho$, the closed-loop trajectories of the system of Eq. 2.1 satisfy:

$$\limsup_{t \to \infty} V\big(x(t)\big) \leq \rho_a.$$

This proves that $x(t) \in \Omega_\rho$ for all times and $x(t)$ is ultimately bounded in Ω_{ρ_a} for the case when $t_{a+1} - t_a = T_m$ for all a and $T_m = N_R \Delta$.

Part 2: In this part, we extend the results proved in Part 1 to the general case, that is, $t_{a+1} - t_a \leq T_m$ for all a and $T_m \leq N_R \Delta$ which implies that $t_{a+1} - t_a \leq N_R \Delta$. Because $f_W(\cdot)$ and $f_V(\cdot)$ are strictly increasing functions of their arguments and $f_V(\cdot)$ is convex, following similar steps as in Part 1, it can be shown that the inequality of Eq. 2.61 still holds. This proves that the stability results stated in Theorem 2.1 hold.   □

Remark 2.6 Theorem 2.1 is important from an MPC point of view because if the maximum time without data losses is smaller than the maximum time that the system can operate in open-loop without leaving the stability region, the feasibility of the optimization problem for all times is guaranteed, since each time feedback is regained, the state is guaranteed to be inside the stability region, thereby yielding a feasible optimization problem.

Remark 2.7 In the LMPC of Eqs. 2.23–2.28, no state constraint has been considered but the presented approach can be extended to handle state constraints by restricting the closed-loop stability region further to satisfy the state constraints.

Remark 2.8 It is also important to remark that when there are data losses in the control system, standard MPC formulations do not provide guaranteed closed-loop stability results. For any MPC scheme, in order to obtain guaranteed closed-loop stability results, even in the case where initial feasibility of the optimization problem is given, the formulation of the optimization problem has to be modified accordingly to take into account data losses in an explicit way.

Remark 2.9 Although the proof of Theorem 2.2 is constructive, the constants obtained are conservative. This is the case with most of the results of the type presented in this book. In practice, the different constants are better estimated through closed-loop simulations. The various inequalities provided are more useful as guidelines on the interaction between the various parameters that define the system and the controller and may be used as guidelines to design the controller and the network.

2.7.4 Application to a Chemical Reactor

Consider a well mixed, nonisothermal continuously stirred tank reactor (CSTR) where three parallel irreversible elementary exothermic reactions take place of the form $A \rightarrow B$, $A \rightarrow C$ and $A \rightarrow D$. B is the desired product and C and D are byproducts. The feed to the reactor consists of pure A at flow rate F, temperature T_{A0} and molar concentration $C_{A0} + \Delta C_{A0}$ where ΔC_{A0} is an unknown time-varying uncertainty. Due to the nonisothermal nature of the reactor, a jacket is used to remove/provide heat to the reactor. Using first principles and standard modeling assumptions, the following mathematical model of the process is obtained [21]:

$$\frac{dT}{dt} = \frac{F}{V_r}(T_{A0} - T) - \sum_{i=1}^{3} \frac{\Delta H_i}{\sigma c_p} k_{i0} e^{\frac{-E_i}{RT}} C_A + \frac{Q}{\sigma c_p V_r}, \qquad (2.62)$$

$$\frac{dC_A}{dt} = \frac{F}{V_r}(C_{A0} + \Delta C_{A0} - C_A) + \sum_{i=1}^{3} k_{i0} e^{\frac{-E_i}{RT}} C_A, \qquad (2.63)$$

where C_A denotes the concentration of the reactant A, T denotes the temperature of the reactor, Q denotes the rate of heat input/removal, V_r denotes the volume of the reactor, $\Delta H_i, k_{i0}, E_i, i = 1, 2, 3$ denote the enthalpies, preexponential constants and activation energies of the three reactions, respectively, and c_p and σ denote the heat capacity and the density of the fluid in the reactor, respectively. The values of the process parameters are shown in Table 2.1.

For $Q_s = 0$ KJ/h (Q_s is the steady-state value of Q), the CSTR of Eqs. 2.62–2.63 has three steady-states (two locally asymptotically stable and one unstable). The control objective is to stabilize the system at the open-loop unstable steady state $T_s = 388$ K, $C_{As} = 3.59$ mol/l. The manipulated input is the rate of heat input Q. We consider a time-varying uncertainty in the concentration of the inflow $|\Delta C_{A0}| \leq 0.5$ kmol/m^3. The control system is subject to data losses in both the sensor-controller and the controller-actuator links.

To demonstrate the theoretical results, we first design the nonlinear control law $h(x)$ as a Lyapunov-based feedback law using the method presented in [97]. The

Table 2.1 Process parameters of the CSTR of Eqs. 2.62–2.63

F	4.998 [m^3/h]	k_{10}	3×10^6 [h^{-1}]
V_r	1 [m^3]	k_{20}	3×10^5 [h^{-1}]
R	8.314 [KJ/kmol K]	k_{30}	3×10^5 [h^{-1}]
T_{A0}	300 [K]	E_1	5×10^4 [KJ/kmol]
C_{A0}	4 [kmol/m^3]	E_2	7.53×10^4 [KJ/kmol]
ΔH_1	-5.0×10^4 [KJ/kmol]	E_3	7.53×10^4 [KJ/kmol]
ΔH_2	-5.2×10^4 [KJ/kmol]	σ	1000 [kg/m^3]
ΔH_3	-5.4×10^4 [KJ/kmol]	c_p	0.231 [KJ/kg K]

CSTR of Eqs. 2.62–2.63 belongs to the following class of nonlinear systems:

$$\dot{x}(t) = f\big(x(t)\big) + g\big(x(t)\big)u(t) + w\big(x(t)\big), \tag{2.64}$$

where $x^T = [T - T_s \ C_A - C_{As}]$ is the state, $u = Q - Q_s$ is the input and $w = \Delta C_{A0}$ is a time varying bounded disturbance with the upper bound $\theta = 0.5$ kmol/m^3. We consider the Lyapunov function $V(x) = x^T P x$ with:

$$P = \begin{bmatrix} 1 & 0 \\ 0 & 10^4 \end{bmatrix}. \tag{2.65}$$

The values of the weights have been chosen to account for the different range of numerical values for each state. The following feedback law [97] asymptotically stabilizes the open-loop unstable steady-state of the nominal process:

$$h(x) = \begin{cases} -\dfrac{L_f V + \sqrt{(L_f V)^2 + (L_g V)^4}}{L_g V} & \text{if } L_g V \neq 0, \\ 0 & \text{if } L_g V = 0, \end{cases} \tag{2.66}$$

where $L_f V = \frac{\partial V(x)}{\partial x} f(x)$ and $L_g V = \frac{\partial V(x)}{\partial x} g(x)$ denote the Lie derivatives of the scalar function V with respect to the vectors fields f and g in Eq. 2.64, respectively. This controller will be used in the design of the LMPC of Eqs. 2.16–2.20 and the LMPC of Eqs. 2.23–2.28. The stability region Ω_ρ is defined as $V(x) \leq 1000$, i.e., $\rho = 1000$.

First, we have to choose an appropriate sampling time and a maximum prediction horizon for the LMPC based on the properties of $h(x)$. The inequalities obtained in the main results of this section are conservative to be used to estimate an appropriate sampling time for a given uncertainty bound and the maximum time that the system can operate in open-loop without leaving the stability region. In order to obtain practical estimates, we resort to extensive off-line closed-loop simulations under the Lyapunov-based controller of Eq. 2.66. After trying different sampling times, we choose $\Delta = 0.05$ h. For this sampling time, the closed-loop system with $u = h(x)$ is practically stable and the performance is similar to the closed-loop system with continuous measurements. With this sampling time, the maximum time such that the system remains in Ω_ρ when controlled in open-loop with the nominal sampled input trajectory is 5Δ (i.e., $N_R = 5$). This value is also estimated using data from simulations.

We implement the LMPCs presented in the previous sections using a sampling time $\Delta = 0.05$ h and a prediction horizon $N = N_R = 5$. The cost function is defined by the weighting matrices $Q_c = P$ and $R_c = 10^{-6}$. The values of the weights have been tuned in a way such that the values of the control inputs are comparable to the ones computed by the Lyapunov-based controller (i.e., same order of magnitude of the input signal and convergence time of the closed-loop system when no uncertainty or data losses are taken into account).

We will first compare the LMPC of Eqs. 2.23–2.28 with the original LMPC of Eqs. 2.16–2.20. In this scheme, no data losses were taken into account. We implement the two LMPCs using the same strategy, that is, sending to the actuator the

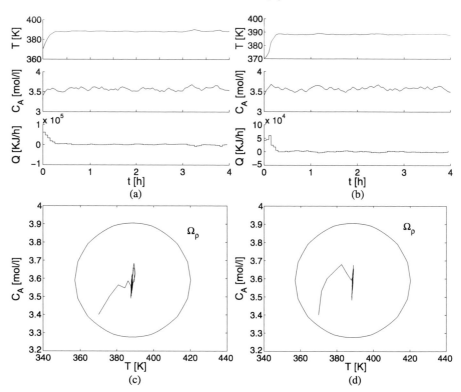

Fig. 2.2 (**a, c**) State and input trajectories of the CSTR of Eqs. 2.62–2.63 with the LMPC of Eqs. 2.23–2.28 with no data losses; (**b, d**) state and input trajectories of the CSTR of Eqs. 2.62–2.63 with the LMPC of Eqs. 2.16–2.20 with no data losses

whole optimal input trajectory, so in case data losses occur, the input is updated as in the modified receding horizon scheme. The same weights, sampling time and prediction horizon are used.

In Fig. 2.2, the trajectories of both LMPCs are shown assuming no data is lost, that is, the state $x(t_k)$ is available every sampling time. It can be seen that both closed-loop systems are practically stable. Note that regarding optimality, for a given state, the LMPC of Eqs. 2.16–2.20 (not necessarily the closed-loop trajectory) yields a lower cost than the LMPC of Eqs. 2.23–2.28, because the constraints that define the LMPC of Eqs. 2.16–2.20 are less restrictive (i.e., the Lyapunov-based constraint must hold only in the first sampling time whereas in the LMPC of Eqs. 2.23–2.28 it must hold along the whole prediction horizon).

When data losses occur, the LMPC of Eqs. 2.23–2.28 is more robust. The stability region is an invariant set for the closed-loop system if $T_m \leq N\Delta$. That is not the case with the LMPC of Eqs. 2.16–2.20. In Fig. 2.3, the trajectories of the closed-loop system under both LMPCs are shown for the worst case of data loss scenario with $T_m = 5\Delta$; that is, the system receives only one measurement of the actual state every 5 samples. These trajectories account for the worst-case effect

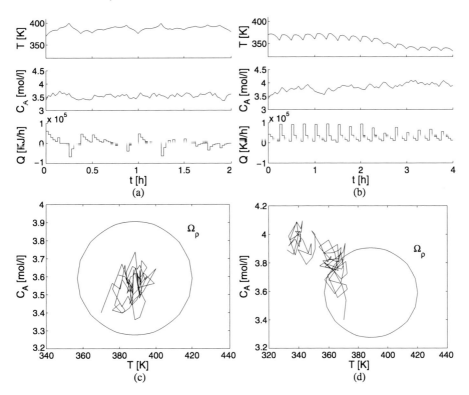

Fig. 2.3 (**a, c**) Worst case state and input trajectories of the CSTR of Eqs. 2.62–2.63 with the LMPC of Eqs. 2.23–2.28 with $T_m = 5\Delta$; (**b, d**) state and input trajectories of the CSTR of Eqs. 2.62–2.63 with the LMPC of Eqs. 2.16–2.20 with $T_m = 5\Delta$

of the data losses. The trajectories are shown in the state space along with the closed-loop stability region Ω_ρ. It can be seen that the trajectory under the LMPC of Eqs. 2.16–2.20 leaves the stability region, while the trajectory under the LMPC of Eqs. 2.23–2.28 remains inside. When data losses are taken into account, in order to inherit the stability properties of the Lyapunov-based controller of Eq. 2.66, the constraints must be modified to take into account data losses as in the LMPC of Eqs. 2.23–2.28.

We now compare the LMPC of Eqs. 2.23–2.28 with the Lyapunov-based controller of Eq. 2.66 applied in a sample-and-hold fashion following a "last available control" strategy, i.e., when data is lost, the actuator keeps implementing the last received input value. Note that, through extensive simulations, we have found that in this particular example, the strategy of setting the input to zero when data losses occur, yields worst results than the strategy of implementing the last available input. In Fig. 2.4, the worst case trajectories with $T_m = 2\Delta$ for both controllers are shown. It can be seen that, due to the instability of the open-loop steady state, for this small amount of losses, the Lyapunov-based controller is not able to stabilize the system.

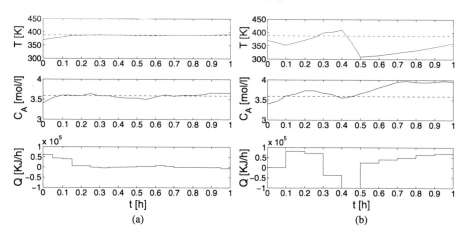

Fig. 2.4 Worst case state and input trajectories of the CSTR of Eqs. 2.62–2.63 with $T_m = 2\Delta$ in closed-loop with (**a**) the LMPC of Eqs. 2.23–2.28 and (**b**) the Lyapunov-based controller of Eq. 2.66

Table 2.2 Total performance costs along the closed-loop trajectories of the CSTR of Eqs. 2.62–2.63 under the Lyapunov-based controller of Eq. 2.66 and the LMPC of Eqs. 2.23–2.28

sim.	Lyapunov-based controller of Eq. 2.66	LMPC of Eqs. 2.23–2.28
1	0.1262×10^{12}	0.0396×10^{12}
2	0.3081×10^{12}	0.2723×10^{12}
3	0.0561×10^{12}	0.0076×10^{12}
4	0.9622×10^{11}	0.2884×10^{11}
5	3.8176×10^{11}	1.3052×10^{11}
6	0.9078×10^{11}	0.0950×10^{11}
7	0.4531×10^{12}	0.2678×10^{12}
8	0.6752×10^{11}	0.5689×10^{11}
9	1.0561×10^{11}	0.6776×10^{11}
10	0.5332×10^{12}	0.3459×10^{12}

This is due to the fact that this control scheme does not update the control actuator output using the model, as the LMPC of Eqs. 2.23–2.28 does.

We have also carried out another set of simulations to demonstrate that the LMPC of Eqs. 2.23–2.28, although inherits the same stability and robustness properties of the Lyapunov-based controller that it employs, it does outperform the Lyapunov-based controller of Eq. 2.66 from a performance index point of view. Table 2.2 shows the total cost computed for 10 different closed-loop simulations under the LMPC and the Lyapunov-based controller implemented in a sample-and-hold fashion, using the nominal model to predict the evolution of the system when data is lost. To carry out this comparison, we compute the total cost of each simulation based on

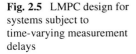

Fig. 2.5 LMPC design for systems subject to time-varying measurement delays

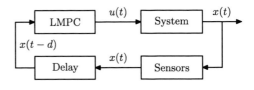

the performance index of the LMPC which has the form:

$$\int_{t_0}^{t_f} \left[\left\| x(\tau) \right\|_{Q_c} + \left\| u(\tau) \right\|_{R_c} \right] d\tau, \tag{2.67}$$

where $t_0 = 0$ is the initial time of the simulations and $t_f = 4$ h is the end of the simulation. For each pair of simulations (one for each controller), a different initial state inside the stability region, a different random uncertainty trajectory and a different data losses realization is chosen. As it can be seen in Table 2.2, the total cost under the LMPC of Eqs. 2.23–2.28 is lower than the corresponding total cost under the Lyapunov-based controller. This demonstrates that in this example, the LMPC shares the same robustness and stability properties and is more optimal than the Lyapunov-based controller, which is not designed taking into account any optimality consideration.

The simulations have been done in MATLAB® using *fmincon* and a Runge–Kutta solver with a fixed integration time of 0.001 h. To simulate the time-varying uncertainty, a different random value $w(t)$ has been applied at each integration step.

2.8 LMPC with Delayed Measurements

In this section, we deal with the design of LMPC for nonlinear systems subject to time-varying measurement delays in the feedback loop. In the LMPC design that will be presented, when measurement delays occur, the nominal model of the system is used together with the latest available measurement to estimate the current state, and the resulting estimate is used to evaluate the LMPC; at time instants where no measurements are available due to the delay, the actuator implements the last optimal input trajectory evaluated by the controller as discussed in the previous section. The LMPC accounting for delays is also designed based on a nonlinear control law which is able to stabilize the closed-loop system and inherits the stability and robustness properties in the presence of uncertainty and time-varying delays of the nonlinear control law, while taking into account optimality considerations. The closed-loop system considered in this section is shown in Fig. 2.5.

2.8.1 Modeling of Delayed Measurements

We assume that the state of the system of Eq. 2.1 is received by the controller at asynchronous time instants t_a where $\{t_{a\geq0}\}$ is a random increasing sequence of times

Fig. 2.6 A possible sequence
of delayed measurements

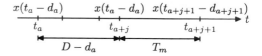

and that there exists an upper bound T_m on the interval between two successive measurements as described in Eq. 2.22. We also assume that there are delays in the measurements received by the controller due to delays in the sampling process and data transmission. In order to model delays in measurements, another auxiliary variable d_a is introduced to indicate the delay corresponding to the measurement received at time t_a, that is, at time t_a, the measurement $x(t_a - d_a)$ is received. In general, if the sequence $\{d_{a\geq0}\}$ is modeled using a random process, there exists the possibility of arbitrarily large delays. In this case, it is improper to use all the delayed measurements to estimate the current state and decide the control inputs, because when the delays are too large, they may introduce enough errors to destroy the stability of the closed-loop system. In order to study the stability properties in a deterministic framework, we assume that the delays associated with the measurements are smaller than an upper bound D, that is:

$$d_a \leq D. \tag{2.68}$$

The size of D is, in general, related to measurement sensor delays and data transmission network delays. We note that for chemical processes, the delay in the measurements received by a controller are mainly caused in the measurement sampling process. We also assume that the time instant when a measurement is sampled is recorded and transmitted together with the measurement. This assumption is practical for many process control applications and implies that the delay in a measurement received by the controller is calculable and can be assumed to be known.

Note that because the delays are time-varying, it is possible that at a time instant t_a, the controller may receive a measurement $x(t_a - d_a)$ which does not provide new information (i.e., $t_a - d_a \leq t_{a-1} - d_{a-1}$); that is, the controller has already received a measurement of the state after time $t_a - d_a$. We assume that each measurement is time-labeled, and hence the controller is able to discard a newly received measurement if $t_a - d_a < t_{a-1} - d_{a-1}$. Figure 2.6 shows part of a possible sequence of $\{t_{a\geq0}\}$. At time t_a, the state measurement $x(t_a - d_a)$ is received. There exists a possibility that between t_a and t_{a+j}, with $t_{a+j} - t_a = D - d_a$ and j being an unknown integer, all the measurements received do not provide new information. Note that any measurements received after t_{a+j} provide new information because the maximum delay is D and the latest received measurement was $x(t_a - d_a)$. The maximum possible time interval between t_{a+j} and t_{a+j+1} is T_m. Therefore, the maximum amount of time in which the system might operate in open-loop following t_a is $D + T_m - d_a$. This upper bound will be used in the formulation of the LMPC design for systems subject to delayed measurements below.

Remark 2.10 The sequences $\{t_{a\geq0}\}$ and $\{d_{a\geq0}\}$ characterize the time needed to obtain a new measurement in the case of asynchronous measurements or the quality of the network link in the case of networked (wired or wireless) communications subject to data losses and time-varying delays. The model is general and can be used to model a wide class of systems subject to asynchronous, delayed measurements.

2.8.2 LMPC Formulation with Measurement Delays

A controller for a system subject to time-varying measurement delays must take into account two important issues. First, when a new measurement is received, this measurement may not correspond to the current state of the system. This implies that in this case, the controller has to make a decision using an estimate of the current state. Second, because the delays are time-varying, the controller may not receive new information every sampling time. This implies that in this case, the controller has to operate in open-loop using the last received measurements. To this end, when a delayed measurement is received the controller uses the nominal system model and the input trajectory that has been applied to the system to get an estimate of the current state and then an MPC optimization problem is solved in order to decide the optimal future input trajectory that will be applied until new measurements are received. This approach implies that the previous control input trajectory should be stored in the controller. The implementation strategy for the LMPC for systems subject to time-varying measurement delays is as follows:

1. When a measurement $x(t_a - d_a)$ is available at t_a, the LMPC checks whether the measurement provides new information. If $t_a - d_a > \max_{l<a} t_l - d_l$, go to Step 2. Else the measurement does not contain new information and is discarded, go to Step 5.
2. The LMPC estimates the current state of the system $\tilde{x}(t_a)$ and computes the optimal input trajectory of u based on $\tilde{x}(t_a)$ for $t \in [t_a, t_a + N\Delta)$.
3. The LMPC sends the entire optimal input trajectory to the actuators.
4. The actuators implement the input trajectory until a new measurement is received at time t_{a+1}.
5. When a new measurement is received ($a \leftarrow a + 1$), go to Step 1.

The LMPC that takes into account time-varying measurement delay in an explicit way is based on the following constrained optimal control problem:

$$\min_{u \in S(\Delta)} \int_{t_a}^{t_a + N\Delta} \left[\|\tilde{x}(\tau)\|_{Q_c} + \|u(\tau)\|_{R_c} \right] d\tau, \tag{2.69}$$

$$\text{s.t.} \quad \dot{\tilde{x}}(t) = f\big(\tilde{x}(t), u(t), 0\big), \quad \forall t \in [t_a - d_a, t_a + N\Delta), \tag{2.70}$$

$$u(t) = u_d^*(t), \quad \forall t \in [t_a - d_a, t_a), \tag{2.71}$$

$$\tilde{x}(t_a - d_a) = x(t_a - d_a), \tag{2.72}$$

$$\dot{\hat{x}}(t) = f\big(\hat{x}(t), h\big(\hat{x}(t_a + j\Delta)\big), 0\big), \quad t \in \big[t_a + j\Delta, t_a + (j+1)\Delta\big), \tag{2.73}$$

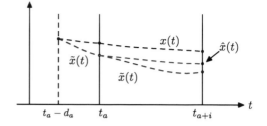

$$\hat{x}(t_a) = \tilde{x}(t_a), \tag{2.74}$$

$$V\big(\tilde{x}(t)\big) \le V\big(\hat{x}(t)\big), \quad \forall t \in [t_a, t_a + N_{D,a}\Delta), \tag{2.75}$$

where $u_d^*(t)$ indicates the actual control input trajectory that has been applied to the system, $x(t_a - d_a)$ is the delayed measurement that is received at t_a with delay size d_a, $\tilde{x}(t_a)$ is an estimate of the current system state, $j = 0, \ldots, N - 1$, and $N_{D,a}$ is the smallest integer satisfying $N_{D,a}\Delta \ge T_m + D - d_a$.

The optimal solution to the LMPC optimization problem of Eqs. 2.69–2.75 is denoted as $u_d^*(t|t_a)$ which is defined for $t \in [t_a, t_a + N\Delta)$. The manipulated input of the system of Eq. 2.1 under the control of the LMPC of Eqs. 2.23–2.28 is defined as follows:

$$u(t) = u_d^*(t|t_a), \quad \forall t \in [t_a, t_{a+i}), \tag{2.76}$$

for all t_a such that $t_a - d_a > \max_{l<a} t_l - d_l$ and for a given t_a, the variable i denotes the smallest integer that satisfies $t_{a+i} - d_{a+i} > t_a - d_a$.

In the LMPC design of Eqs. 2.69–2.75, if at a sampling time, a new measurement $x(t_a - d_a)$ is received, an estimate of the current state $\tilde{x}(t_a)$ is obtained using the nominal model of the system (the constraint of Eq. 2.70) and the control input trajectory applied to the system from $t_a - d_a$ to t_a (the constraint of Eq. 2.71) with the initial condition $\tilde{x}(t_a - d_a) = x(t_a - d_a)$ (the constraint of Eq. 2.72). The estimated state $\tilde{x}(t_a)$ is then used to obtain the optimal future control input trajectory. The LMPC of Eqs. 2.69–2.75 uses the nominal model to predict the future trajectory $\tilde{x}(t)$ for a given input trajectory $u(t) \in S(\Delta)$ with $t \in [t_a, t_a + N\Delta)$. A cost function is minimized (Eq. 2.69), while assuring that the value of the Lyapunov function along the predicted trajectory $\tilde{x}(t)$ satisfies a Lyapunov-based constraint (the constraint of Eq. 2.75) where $\hat{x}(t)$ is the state trajectory corresponding to the nominal system in closed-loop with the nonlinear control law $h(x)$ (the constraint of Eq. 2.73) with the initial condition $\hat{x}(t_a) = \tilde{x}(t_a)$ (the constraint of Eq. 2.74). Note that the length of the constraint $N_{D,a}$ depends on the current delay d_a so it may have different values at different time instants and has to be updated before solving the optimization problem of Eqs. 2.69–2.75. If the controller does not receive any new measurement at a sampling time, it keeps implementing the last evaluated optimal trajectory. This strategy is a receding horizon scheme, which takes time-varying measurement delays explicitly into account.

Figure 2.7 shows a possible scenario for a system of dimension 1. A delayed measurement $x(t_a - d_a)$ is received at time t_a and the next new measurement is not obtained until t_{a+i}. This implies that at time t_a we evaluate the LMPC of Eqs. 2.69–2.75 and we apply the optimal input $u_d^*(t|t_a)$ from t_a to t_{a+i}. The solid vertical lines are used to indicate sampling times in which a new measurement is obtained (that is, t_a and t_{a+i}) and the dashed vertical line is used to indicate the time corresponding to the measurement obtained in t_a (that is, $t_a - d_a$).

2.8.3   Stability Properties

In this subsection, we present the stability properties of the LMPC of Eqs. 2.69–2.75 for systems subject to time-varying measurement delays. Theorem 2.2 below provides sufficient conditions under which the LMPC of Eqs. 2.69–2.75 guarantees stability of the closed-loop system in the presence of time-varying measurement delays.

Theorem 2.2 *Consider the system of Eq. 2.1 in closed-loop, which closes at asynchronous time instants $\{t_{a\geq0}\}$ that satisfy the condition of Eq. 2.22, under the LMPC of Eqs. 2.69–2.75 based on a controller $h(x)$ that satisfies the conditions of Eqs. 2.4–2.7. Let $\Delta, \varepsilon_s > 0, \rho > \rho_{\min} > 0, \rho > \rho_s > 0, N \geq 1$ and $D \geq 0$ satisfy the condition of Eq. 2.31 and the following inequality:*

$$-N_R \varepsilon_s + f_V\big(f_W(N_D \Delta)\big) + f_V\big(f_W(D)\big) < 0. \tag{2.77}$$

with $f_V(\cdot)$ and $f_W(\cdot)$ defined in Eqs. 2.49 and 2.43, respectively, N_D being the smallest integer satisfying $N_D \Delta \geq T_m + D$, and N_R being the smallest integer satisfying $N_R \Delta \geq T_m$. If $N \geq N_D$, $x(t_0) \in \Omega_\rho$ and $d_0 = 0$, then $x(t)$ is ultimately bounded in $\Omega_{\rho_d} \subseteq \Omega_\rho$ where:

$$\rho_d = \rho_{\min} + f_V\big(f_W(N_D \Delta)\big) + f_V\big(f_W(D)\big). \tag{2.78}$$

Proof In order to prove that the system of Eq. 2.1 in closed-loop with the LMPC of Eq. 2.69–2.75 is ultimately bounded in a region that contains the origin, we will prove that the Lyapunov function $V(x)$ is a decreasing function of time with a lower bound on its magnitude. We assume that the delayed measurement $x(t_a - d_a)$ is received at time t_a and that a new measurement is not obtained until t_{a+i}. The LMPC of Eq. 2.69–2.75 is solved at t_a and the optimal input trajectory $u_d^*(t|t_a)$ is applied from t_a to t_{a+i}.

Part 1: In this part, we prove that the stability results stated in Theorem 2.2 hold for $t_{a+i} - t_a = N_{D,a}\Delta$ and all $d_a \leq D$.

The trajectory $\hat{x}(t)$ corresponds to the nominal system in closed-loop with the nonlinear control law $u = h(\hat{x})$ implemented in a sample-and-hold fashion with initial condition $\tilde{x}(t_a)$; please see the constraint of Eqs. 2.73 and 2.74. By Proposition 2.1, the following inequality can be obtained:

$$V\big(\hat{x}(t_{a+i})\big) \leq \max\big\{V\big(\hat{x}(t_a)\big) - N_{D,a}\varepsilon_s, \rho_{\min}\big\}. \tag{2.79}$$

The constraint of Eq. 2.75 guarantees that:

$$V\big(\tilde{x}(t)\big) \leq V\big(\hat{x}(t)\big), \quad \forall t \in [t_a, t_a + N_{D,a}\Delta), \tag{2.80}$$

and the constraint of Eq. 2.74 guarantees that $V(\hat{x}(t_a)) = V(\tilde{x}(t_a))$. This implies that:

$$V\big(\tilde{x}(t_{a+i})\big) \leq \max\big\{V\big(\tilde{x}(t_a)\big) - N_{D,a}\varepsilon_s, \rho_{\min}\big\}. \tag{2.81}$$

When $x(t) \in \Omega_\rho$ for all times (this point will be proved below), we can apply Proposition 2.3 to obtain the following inequalities:

$$V\big(\tilde{x}(t_a)\big) \leq V\big(x(t_a)\big) + f_V\big(\|x(t_a) - \tilde{x}(t_a)\|\big), \tag{2.82}$$

$$V\big(x(t_{a+i})\big) \leq V\big(\tilde{x}(t_{a+i})\big) + f_V\big(\|x(t_{a+i}) - \tilde{x}(t_{a+i})\|\big). \tag{2.83}$$

Applying Proposition 2.2, we obtain the following upper bounds on the deviation of $\tilde{x}(t)$ from $x(t)$:

$$\|x(t_a) - \tilde{x}(t_a)\| \leq f_W(d_a), \tag{2.84}$$

$$\|x(t_{a+i}) - \tilde{x}(t_{a+i})\| \leq f_W(N_D\Delta). \tag{2.85}$$

Note that the constraints of Eqs. 2.70–2.72 and the implementation procedure allow us to apply Proposition 2.2 because it is guaranteed that the actual system state $x(t)$ and the state estimated using the nominal model $\tilde{x}(t)$ are obtained using the same input trajectory. Note also that we have taken into account that $N_D\Delta \geq T_m + D - d_a$ for all d_a. Using the inequalities of Eqs. 2.81–2.84, the following upper bound on $V(x(t_{k+j}))$ is obtained:

$$V\big(x(t_{a+i})\big) \leq \max\big\{V\big(x(t_a)\big) - N_{D,a}\varepsilon_s, \rho_{\min}\big\} + f_V\big(f_W(d_a)\big) + f_V\big(f_W(N_D\Delta)\big). \tag{2.86}$$

In order to prove that the Lyapunov function is decreasing between two consecutive new measurements, the following inequality must hold:

$$N_{D,a}\varepsilon_s > f_V\big(f_W(N_D\Delta)\big) + f_V\big(f_W(d_a)\big) \tag{2.87}$$

for all possible $0 \leq d_a \leq D$. Taking into account that $f_W(\cdot)$ and $f_V(\cdot)$ are strictly increasing functions of their arguments, that $N_{D,a}$ is a decreasing function of the delay d_a and that if $d_a = D$ then $N_{D,a} = N_R$, if the condition of Eq. 2.77 is satisfied, the condition of Eq. 2.87 holds for all possible d_a and there exists $\varepsilon_w > 0$ such that the following inequality holds:

$$V\big(x(t_{a+i})\big) \leq \max\big\{V\big(x(t_a)\big) - \varepsilon_w, \rho_d\big\}, \tag{2.88}$$

which implies that if $x(t_a) \in \Omega_\rho/\Omega_{\rho_d}$, then $V(x(t_{a+i})) < V(x(t_a))$, and if $x(t_a) \in \Omega_{\rho_d}$, then $V(x(t_{a+i})) \leq \rho_d$.

Because the upper bound on the difference between the Lyapunov function of the actual trajectory x and the nominal trajectory $\tilde{x}$ is a strictly increasing function of time, the inequality of Eq. 2.88 also implies that:

$$V\left(x(t)\right) \le \max\{V\left(x(t_a)\right), \rho_d\}, \quad \forall t \in [t_a, t_{a+i}). \tag{2.89}$$

Using the inequality of Eq. 2.89 recursively, it can be proved that if $x(t_0) \in \Omega_\rho$, then the closed-loop trajectories of the system of Eq. 2.1 under the LMPC of Eqs. 2.69–2.75 stay in Ω_ρ for all times (i.e., $x(t) \in \Omega_\rho, \forall t$). Moreover, using the inequality of Eq. 2.89 recursively, it can be proved that if $x(t_0) \in \Omega_\rho$, the closed-loop trajectories of the system of Eq. 2.1 under the LMPC of Eqs. 2.69–2.75 satisfy:

$$\limsup_{t \to \infty} V\left(x(t)\right) \le \rho_d. \tag{2.90}$$

This proves that $x(t) \in \Omega_\rho$ for all times and $x(t)$ is ultimately bounded in Ω_{ρ_d} for the case when $t_{a+i} - t_a = N_{D,a}\Delta$.

Part 2: In this part, we extend the results proved in Part 1 to the general case, that is, $t_{a+i} - t_a \le N_{D,a}\Delta$. Taking into account that $f_V(\cdot)$ and $f_W(\cdot)$ are strictly increasing functions of their arguments and $f_V(\cdot)$ is convex, following similar steps as in Part 1, it can be shown that the inequality of Eq. 2.87 holds for all possible $d_a \le D$ and $t_{a+i} - t_a \le N_{D,a}\Delta$. Using this inequality and following the same line of arguments as in the previous part, the stability results stated in Theorem 2.2 can be proved.   □

Remark 2.11 When time-varying measurement delays are not present and new measurements of $x(t)$ are fed into the controller every synchronous sampling time, the LMPC of Eqs. 2.69–2.75 may be simplified to the LMPC of Eqs. 2.16–2.20. Comparing the LMPC of Eqs. 2.16–2.20 with the one of Eqs. 2.69–2.75, the difference is that the Lyapunov-based constraint of Eq. 2.20 has to hold only for one time step. This implies that even if the same implementation procedure is used, and the same optimization problem is solved (in order to estimate the current state), if the Lyapunov-based constraint is not changed, stability cannot be proved. This point will be illustrated in the example in Sect. 2.8.4.

Remark 2.12 In the LMPC of Eqs. 2.23–2.28 for systems with asynchronous feedback without delays, the Lyapunov-based constraint of Eq. 2.28 has to hold for a time period which is equal to or bigger than the maximum time without new measurement. This constraint makes the computed control action more conservative (and thus less optimal) because the controller may have to satisfy the Lyapunov-based constraint over unnecessarily large horizons. If the LMPC of Eqs. 2.23–2.28 is implemented for systems subject to time-varying delays, it will be, in general, less optimal than the LMPC of Eqs. 2.69–2.75. This point will also be illustrated in the example in Sect. 2.8.4.

2.8.4 Application to a Chemical Reactor

Consider the CSTR described by Eqs. 2.62–2.63 in Sect. 2.7.4. We assume that the manipulated input (the rate of heat input Q) is bounded by $|Q| \leq 10^5$ KJ/h and the time-varying uncertainty in the reactant concentration of the inflow is bounded by $|\Delta C_{A0}| \leq 0.2$ mol/l. The control system is subject to time-varying measurement delay in the measurements of the concentration of the reactant, C_A, and in the measurements of the temperature, T. Note that we do not consider the possible different sampling rates of temperature and concentration sensors in this example. Note also that the delay in the measurements could be regarded as the total time needed for online sensors to get a sample, analyze the sample and transmit the data to the controller. The same nonlinear controller of Eq. 2.66 with the same Lyapunov function $V(x)$ and weighting matrix P is used in the design of the LM-PCs used in the simulations. The stability region Ω_ρ is defined as $V(x) \leq 700$, i.e., $\rho = 700$.

The sampling time of the LMPCs is chosen to be $\Delta = 0.025$ h, the maximum allowable measurement delay is $D = 6\Delta = 0.15$ h and the maximum interval between two consecutive measurements is $T_m = \Delta = 0.025$ h which implies that there is a measurement available every Δ but it may not contain new state information. The cost function is defined by the weighting matrices $Q_c = P$ and $R_c = 10^{-6}$.

We first compare the LMPC of Eqs. 2.69–2.75 with the LMPC of Eqs. 2.16–2.20 in the case where no time-varying measurement delays are present. For this simulation, we choose the prediction horizon of the two LMPCs N equal to 7 ($N \geq D + T_m$). We implement the LMPC of Eqs. 2.16–2.20 using the same approach employed in the implementation of the LMPC of Eqs. 2.69–2.75, that is, the current state is estimated using the nominal model when a delayed measurement is received and the last optimal input is applied when no new measurement is received. In Fig 2.8, the trajectories of the CSTR under both LMPCs are shown assuming no measurement delay is present, that is, the state $x(t_k)$ is available every sampling time. It can be seen that both closed-loop systems are practically stable and the trajectories remain in the stability region Ω_ρ.

In order to simulate the process in the presence of measurement delay, we use a random process to generate the delay sequence $\{d_{a \geq 0}\}$, and the time sequence $\{t_{a \geq 0}\}$ and corresponding delay sequence $\{d_{a \geq 0}\}$ in which the control system is subjected to is shown in Fig. 2.9. In this figure, we see the time-varying nature of the measurement delays and the largest delays are equal to the maximum allowable delay $D = 6\Delta = 0.15$ h. Note that when $d_{a+1} = d_a + \Delta$, the controller does not receive any new measurement.

When time-varying measurement delays are present, the LMPC of Eqs. 2.69–2.75 is more robust. The stability region is invariant for the closed-loop system if $D + T_m \leq N\Delta$. This is not the case with the LMPC of Eqs. 2.16–2.20. In Fig. 2.10, the trajectories of the closed-loop system under both controllers are shown in the presence of measurement delay with $D = 6\Delta = 0.15$ h. It can be seen that the

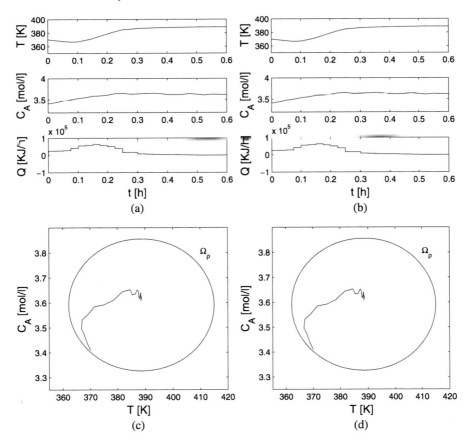

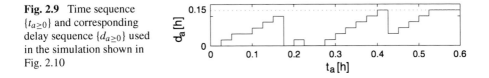

Fig. 2.8 (**a, c**) State and input trajectories of the CSTR of Eqs. 2.62–2.63 with the LMPC of Eqs. 2.69–2.75 when no measurement delay is present; (**b, d**) state and input trajectories of the CSTR of Eqs. 2.62–2.63 with the LMPC of Eqs. 2.16–2.20 when no measurement delay is present

Fig. 2.9 Time sequence $\{t_{a\geq 0}\}$ and corresponding delay sequence $\{d_{a\geq 0}\}$ used in the simulation shown in Fig. 2.10

LMPC of Eqs. 2.16–2.20 can not stabilize the system at the desired open-loop unstable steady-state and the trajectories leave the stability region, while the LMPC of Eqs. 2.69–2.75 keeps the trajectories inside the stability region. When measurement delay is present, in order to provide stability guarantees, the constraints must be modified to take into account the measurement delay as in the LMPC of Eqs. 2.69–2.75.

We have also carried out a set of simulations to compare the LMPC of Eqs. 2.69–2.75 with the LMPC of Eqs. 2.23–2.28 for nonlinear systems subject

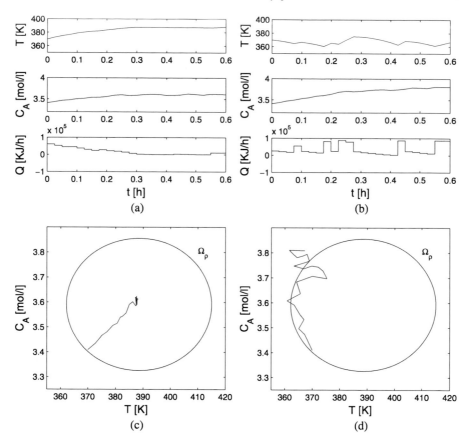

Fig. 2.10 (**a**, **c**) State and input trajectories of the CSTR of Eqs. 2.62–2.63 with the LMPC of Eqs. 2.69–2.75 when D is 6Δ and $T_m = \Delta$; (**b**, **d**) state and input trajectories of the CSTR of Eqs. 2.62–2.63 with the LMPC of Eqs. 2.16–2.20 when D is 6Δ and $T_m = \Delta$

to data losses from a performance index point of view. We also implement the LMPC of Eqs. 2.23–2.28 using the same approach employed in the implementation of the LMPC of Eqs. 2.69–2.75. Table 2.3 shows the total cost computed for 20 different closed-loop simulations under the LMPC of Eqs. 2.69–2.75 and the LMPC of Eqs. 2.23–2.28. To carry out this comparison, we have computed the total cost of each simulation based on the performance index of Eq. 2.67 with the initial simulation time $t_0 = 0$ and the final simulation time $t_f = 2$ h.

The prediction horizon in this set of simulations is $N = 10$. For each pair of simulations (one for each controller) a different initial state inside the stability region, a different uncertainty trajectory and a different random measurement delay sequence is chosen. As can be seen in Table 2.3, the LMPC of Eqs. 2.69–2.75 has a cost lower than the corresponding total cost under the LMPC designed for systems subject to data losses in 16 out of 20 simulations (see also Remark 2.12). This illustrates that the LMPC of Eqs. 2.69–2.75 is, in general, more optimal. This is because the LMPC

Table 2.3 Total performance costs along the closed-loop trajectories of the CSTR of Eqs. 2.62–2.63 under LMPC of Eqs. 2.69–2.75 and LMPC of Eqs. 2.23–2.28	sim.	LMPC of Eqs. 2.69–2.75	LMPC of Eqs. 2.23–2.28
	1	1.8295×10^4	2.4428×10^4
	2	4.2057×10^4	6.0522×10^4
	3	3.2481×10^3	1.0428×10^4
	4	7.4328×10^2	7.3961×10^2
	5	1.4229×10^3	2.7798×10^5
	6	4.9435×10^1	6.1596×10^4
	7	3.2519×10^4	3.4319×10^4
	8	2.7590×10^4	4.7075×10^4
	9	9.4216×10^2	9.4866×10^2
	10	5.4505×10^2	5.4322×10^2
	11	1.9723×10^4	3.1282×10^4
	12	2.7235×10^4	3.8772×10^4
	13	1.8671×10^3	1.9200×10^3
	14	3.7789×10^4	4.0050×10^4
	15	2.1839×10^3	2.1392×10^3
	16	4.2920×10^4	4.4594×10^4
	17	1.5153×10^2	1.7190×10^2
	18	4.9955×10^3	9.9094×10^3
	19	3.2086×10^4	4.8838×10^4
	20	1.5420×10^3	1.5197×10^3

designed for system subject to data losses requires the Lyapunov-based constraint of Eq. 2.28 to be satisfied along the whole possible maximum open-loop operation time (that is $t \in [t_a, t_a + N_R \Delta)$) which yields a more conservative controller from a performance point of view.

We have also carried out a set of simulations to study the dependence on the value of the maximum delay D of the set in which the trajectory of the process under the proposed LMPC scheme is ultimately bounded. In order to estimate the size of each set for a given D, we start the system very close to the equilibrium state and run it for a sufficient long time. In this set of simulations, we set $\Delta C_{A0} = 0.1$ kmol/m^3 and $N = 7$. The simulation time is 25 h. Figure 2.11 shows the location of the states, (C_A, T), at each sampling time and the estimated regions for $D = 2\Delta, 4\Delta, 6\Delta$. Three ellipses are used to estimate the boundaries of the sets, and they are chosen to be as small as possible but still include all the corresponding points indicating the states. From Fig. 2.11, we see that the size of these sets becomes larger as D increases. The results are expected because the size of the sets is not only dependent on the system and the controller, but it also depends on the maximum measurement delay. The longer the size of the delay, the further the system can move away from the steady-state which means a larger set (if the state is still in the stability region Ω_ρ). Note that all the sets for $D = 2\Delta, 4\Delta, 6\Delta$ are included

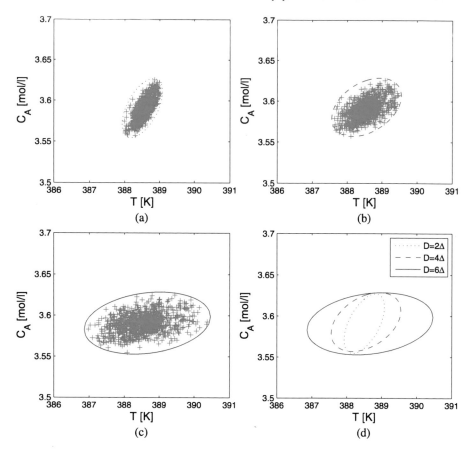

Fig. 2.11 (**a**) Estimate of the set in which the state trajectories of the CSTR of Eqs. 2.62–2.63 with the LMPC of Eqs. 2.69–2.75 are ultimately bounded when the maximum allowable measurement delay D is 2Δ; (**b**) estimate of the set in which the state trajectories of the CSTR of Eqs. 2.62–2.63 with the LMPC of Eqs. 2.69–2.75 are ultimately bounded when the maximum allowable measurement delay D is 4Δ; (**c**) estimate of the set in which the state trajectories of the CSTR of Eqs. 2.62–2.63 with the LMPC of Eqs. 2.69–2.75 are ultimately bounded when the maximum allowable measurement delay D is 6Δ; (**d**) comparison of the three sets

in the stability region of the closed loop system under the LMPC accounting for time-varying delays (Ω_ρ, $\rho = 700$).

2.9 Conclusions

In this chapter, LMPC designs were developed for the control of a broad class of nonlinear uncertain systems subject to data losses/asynchronous measurements and time-varying measurement delays. The main idea is that in order to provide guaranteed stability results in the presence of data losses or time-varying measurement delays, the constraints that define the LMPC optimization problems as well as the implementation procedures have to be modified to account for data losses/asynchronous measurements or time-varying measurement delays. The presented LMPCs possess an explicit characterization of the closed-loop system stability regions. The applications of the presented LMPCs were illustrated using a nonlinear CSTR example.

Chapter 3
Networked Predictive Process Control

3.1 Introduction

In Chap. 2, we presented two LMPC designs for networked control systems subject to feedback data losses and time-varying measurement delays. From a control
system architecture point of view, the two LMPC designs are centralized and aim
to replace existing, dedicated control systems. In this chapter, we present a two-tier
networked control architecture to augment existing, point-to-point control systems
with networked control systems, which take advantage of real-time wired or wireless
sensor and actuator networks. This two-tier control architecture is a decentralized
control architecture and involves the use of hybrid communication networks. In this
case, key issues that need to be carefully handled at the control system design level
include data losses due to field interference, and time-delays due to network traffic
as well as measurement sampling.

The class of networked control problems considered in this chapter arises naturally in the context of process control systems based on hybrid communication
networks (i.e., point-to-point wired links integrated with networked wired or wireless communication) and utilizing multiple heterogeneous measurements (e.g., temperature and concentration). Assuming that there exists a lower-tier control system which relies on point-to-point communication and continuous measurements to
stabilize the closed-loop system, we use LMPC to design an upper-tier networked
control system which profits from both continuous and asynchronous/delayed measurements as well as from additional networked control actuators. The main idea is
to formulate appropriate constraints in the MPC optimization problem based on the
existing lower-tier control system, in a way such that the MPC inherits the robustness and stability properties of the lower-tier controller. The two-tier control system
architecture has the ability to preserve the stability properties of the lower-tier control system while improving the closed-loop performance. The applicability and
effectiveness of the two-tier control architecture is demonstrated using two chemical process examples. Moreover, the two-tier control architecture is also applied to
the optimal management and operation of a standalone hybrid wind–solar energy
generation system. Specifically, we design a supervisory control system via MPC

P.D. Christofides et al., *Networked and Distributed Predictive Control*,
Advances in Industrial Control,
DOI 10.1007/978-0-85729-582-8_3, © Springer-Verlag London Limited 2011

which computes the power references for the wind and solar subsystems at each sampling time while minimizing a suitable cost function. The power references are sent to two local controllers which drive the two subsystems to the requested power references. We explicitly incorporate some important practical considerations, for example, how to extend the life time of the equipment by reducing the peak values of inrush or surge currents, into the formulation of the MPC optimization problem. We present several simulation case studies that demonstrate the applicability and effectiveness of the supervisory predictive control architecture. The results of this chapter were first presented in [51, 57, 88].

3.2 System Description

In this chapter, we consider nonlinear systems described by the following state-space model:

$$\dot{x}(t) = f\big(x(t), u_s(t), u_a(t), w(t)\big), \tag{3.1}$$

$$y_s(t) = h_s\big(x(t)\big), \tag{3.2}$$

$$y_a(t) = h_a\big(x(t)\big), \tag{3.3}$$

where $x(t) \in R^n$ denotes the vector of state variables, $y_s(t) \in R^{n_s}$ denotes measurements that are available continuously, $y_a(t) \in R^{n_a}$ denotes measurements that are sampled at asynchronous time instants, $u_s(t) \in R^{m_s}$ and $u_a(t) \in R^{m_a}$ are two different sets of possible control inputs, and $w(t) \in R^w$ denotes the vector of disturbance variables. The disturbance vector is assumed to be bounded, i.e., $w(t) \in W$ where:

$$W := \big\{w \in R^w : \|w\| \le \theta, \theta > 0\big\} \tag{3.4}$$

with θ being a known positive real number.

We assume that f is a locally Lipschitz vector function, h_s and h_a are sufficiently smooth vector functions, $f(0,0,0,0) = 0$, $h_s(0) = 0$ and $h_a(0) = 0$. This means that the origin is an equilibrium point for the nominal system with $u_s = 0$ and $u_a = 0$.

The system of Eqs. 3.1–3.3 has both continuous synchronous and sampled asynchronous measurements. We assume that $y_s(t)$ is available for all t, while $y_a(t)$ is sampled and only available at some time instants t_a where $\{t_{a\geq0}\}$ is a random increasing sequence of times. Moreover, there may be time-varying measurement delays associated with the asynchronous measurements $y_a(t)$. Please see Sect. 3.3 for a precise definition of the measurement/network model that considered in this chapter.

Remark 3.1 The two sets of inputs include both systems with multiple inputs, or systems with a single input divided artificially into two parts; that is:

$$\dot{x}(t) = \hat{f}\big(x(t), u(t), w(t)\big) \tag{3.5}$$

with $u(t) = u_s(t) + u_a(t)$. This implies that the two-tier control architecture presented in this chapter can be used to design control systems which produce adjustments to the actions of an already operating local control system to improve the closed-loop performance.

3.3 Modeling of Measurements

The system of Eqs. 3.1–3.3 is controlled using both continuous synchronous, y_s, and asynchronous, delayed measurements, y_a. This class of systems arises naturally in process control applications, where different process variables have to be measured such as temperature, flow rates, species concentrations or particle size distributions. This model is also of interest in the context of processes controlled through a hybrid communication network in which networked wired/wireless sensors and actuators are used to add redundancy to existing control loops (which use point-to-point wired communication links and continuous measurements) because networked communication is often subject to data losses due to field interference (for example, in wireless communication) and time-varying delays due to network traffic.

We assume that y_s is available for all t, while delayed y_a samples are received at an asynchronous rate. We also assume that each y_a measurement is time-labeled, so the controller is able to discard nonrelevant information. Delays in the computation and implementation of control actions can be readily lumped with the measurement delays and are not treated separately. The time instants at which a new delayed y_a sample is received are denoted t_a, where $\{t_{a\geq 0}\}$ is a random increasing sequence of times. To model the time-varying delay, an auxiliary variable d_a is introduced to indicate the delay corresponding to the sample received at time t_a, that is, at time instant t_a, the sample $y_a(t_a - d_a) = h_a(x(t_a - d_a))$ is received.

In general, if the sequence $\{d_{a\geq 0}\}$ is modeled using a random process, it is improper to use all the delayed measurements to estimate the current state and decide the control inputs, because when the delays are too large, they may introduce enough errors to destroy the stability of the closed-loop system. In order to study the stability properties in a deterministic framework, in this chapter, we only take advantage of delayed measurements such that the delays associated with the measurements are smaller than an upper bound D, i.e., $d_a \leq D, a = 0, 1 \ldots$. The sequence $\{t_{a\geq 0}\}$ only indicates time instants in which new measurements are available with a corresponding measurement delay smaller than or equal to D.

We assume that the measurement of the full state x can be obtained by a proper combination of measurements y_s and y_a at a given time instant. Due to the asynchronous nature of y_a, the time interval between two consecutive state x measurements is unknown, moreover, due to the time-varying measurement delay of y_a, the full state x is also subject to time-varying delays. This implies that a controller that is designed to profit from the extra information provided by the asynchronous, delayed measurements y_a must take into account that between two consecutive state measurements it has to operate in open-loop and that the received state measurements

Fig. 3.1 Lower-tier
controller with dedicated
point-to-point, wired
communication links and
continuous sensing and
actuation

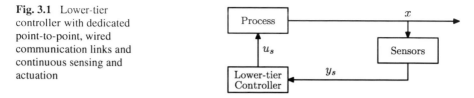

are delayed so the real state of the system has to be estimated using the nominal
model of the system and the available measurement information.

Remark 3.2 The sequence $\{t_{a\geq0}\}$ does not take into account time instants in which
a sample that does not provide new information or a sample that involves a delay
larger than D is received, that is, the controller discards samples with already known
information, or with a delay too large to use this sample to estimate the current state
(recall that the measurements are time-labeled).

Remark 3.3 We have considered that the delayed full state is available asynchronously to simplify the notation. The results can be extended to controllers based
on partial state information.

3.4 Lower-Tier Controller

The continuous measurements $y_s(t)$ can be used to design a continuous output feedback controller to stabilize the system. We term the control system based only on
the continuous measurements $y_s(t)$ as lower-tier controller. This controller does not
use the asynchronous measurements $y_a(t)$. Figure 3.1 shows a schematic of the
lower-tier control system. Following this idea, we assume that there exists an output feedback controller $u_s(t) = k_s(y_s)$ (where $k_s(y_s)$ is assumed to be a sufficiently
smooth function of y_s) that renders the origin of the nominal closed-loop system
asymptotically stable with $u_a(t) \equiv 0$. Using converse Lyapunov theorems [11, 40,
48, 64], this assumption implies that there exist functions $\alpha_i(\cdot)$, $i = 1, 2, 3, 4$ of
class $\mathcal{K}$ and a continuously differentiable Lyapunov function $V(x)$ for the nominal
closed-loop system, that satisfy the following inequalities:

$$\alpha_1(\|x\|) \leq V(x) \leq \alpha_2(\|x\|), \tag{3.6}$$

$$\frac{\partial V(x)}{\partial x} f(x, k_s(h_s(x)), 0, 0) \leq -\alpha_3(\|x\|), \tag{3.7}$$

$$\left\| \frac{\partial V(x)}{\partial x} \right\| \leq \alpha_4(\|x\|), \tag{3.8}$$

for all $x \in O \subseteq R^n$ where O is an open neighborhood of the origin. We denote
the region $\Omega_\rho \subseteq O$ as the stability region of the closed-loop system under the controller $k_s(y_s)$. In the remainder, we will refer to the controller $k_s(y_s)$ as the lower-tier
controller.

The lower-tier controller $k_s(y_s)$ is able to stabilize the system, however, it does not profit from the extra information provided by $y_a(t)$. In the remainder of this chapter, we present a two-tier control architecture that profits from this extra information to improve closed-loop performance.

Remark 3.4 The assumption that there exists a lower-tier controller which can stabilize the closed-loop system using only the continuous measurements $y_s(t)$ and the inputs $u_s(t)$ implies that, in principle, it is not necessary to use the additional information provided by the asynchronous measurements and the extra inputs $u_a(t)$ in order to achieve closed-loop stability. However, the main objective of the two-tier control architecture is to profit from this extra information and control effort to improve the closed-loop performance while maintaining the stability properties achieved by the lower-tier controller.

Remark 3.5 Note that in many application areas, specifically in chemical plants, there are control systems that have already been implemented using dedicated, local control networks. These control systems will not be replaced by networked control systems. Instead, networked control systems should be designed and implemented to augment the preexisting control systems to maintain stability and improve closed-loop performance. This is why we assume that there exists a preexisting stabilizing controller $k_s(y_s)$ for the lower-tier control system based on the continuous measurements $y_s(t)$.

Remark 3.6 We have considered static lower-tier controllers to simplify the notation. The formulation can be extended to dynamic lower-tier controllers. In the examples in Sects. 3.5.3, 3.5.4, 3.6.3 and 3.6.4, proportional-integral (PI) controllers are used as the lower-tier controllers.

Remark 3.7 The lower-tier controller provides some degree of robustness with respect to the uncertainty w. The conditions of Eqs. 3.6–3.8 and the Lipschitz property of f guarantee that: (a) the closed-loop nominal system under the lower-tier controller is asymptotically stable; (b) the closed-loop system state under the lower-tier controller subject to the disturbances is ultimately bounded, provided θ is sufficiently small, in a region that contains the origin that depends on the size of the uncertainty. These properties are made explicit in Proposition 3.1 in the next section. Please see [40] for more details.

3.5   Two-Tier Networked Control Architecture with Continuous/Asynchronous Measurements

In this section, we consider the design of the two-tier control architecture for the system of Eqs. 3.1–3.3 with continuous and asynchronous measurements without delays; that is $d_a = 0$ for all time instants. The extension of the two-tier control

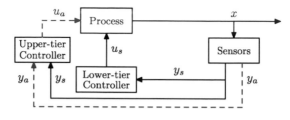

Fig. 3.2 Two-tier networked control architecture (*solid lines* denote dedicated point-to-point, wired communication links and continuous sensing and actuation; *dashed lines* denote networked (wired or wireless) communication or asynchronous sampling and actuation)

architecture for the system of Eqs. 3.1–3.3 with continuous and asynchronous measurements involving time-varying delays (i.e., $d_a \neq 0$) will be presented in Sect. 3.6.

The main objective of the two-tier control architecture is to improve the performance of the closed-loop system using the information provided by $y_a(t)$ while guaranteeing that the stability properties of the lower-tier controller are maintained. This is done by defining a controller (upper-tier controller) based on the full state measurements obtained from both the synchronous and asynchronous measurements at time steps t_a. In the two-tier control architecture, the upper-tier controller decides the trajectory of $u_a(t)$ between successive samples, i.e., for $t \in [t_a, t_{a+1})$ and the lower-tier controller decides $u_s(t)$ using the continuously available measurements. Figure 3.2 shows a schematic of the two-tier control architecture. Due to the asynchronous nature of $y_a(t)$, the upper-tier controller has to take into account that the time interval between two consecutive samples is unknown and there exists the possibility of an infinitely large interval.

Remark 3.8 Note that since the lower-tier controller has already been designed, this controller views the input $u_a(t)$ as a disturbance that has to be rejected if the controller that is used to manipulate $u_a(t)$ is not properly designed. Therefore, the design of the upper-tier controller has to take into account the decisions that will be made by the lower-tier controller to maintain closed-loop stability and guarantee improved closed-loop performance.

3.5.1 Upper-Tier Networked LMPC Formulation

In order to take advantage of the model of the system and the asynchronous state measurements, we use MPC to decide $u_a(t)$. The main idea is the following: at each time instant t_a that a new state measurement is obtained, an open-loop finite horizon optimal control problem is solved and an optimal input trajectory is obtained. This input trajectory is implemented until a new measurement arrives at time t_{a+1}. If the time between two consecutive measurements is longer than the prediction horizon, $u_a(t)$ is set to zero until a new measurement arrives and the optimal control problem is solved again. In order to guarantee that the resulting closed-loop system is stable, we design the MPC via LMPC. In the LMPC designs presented in Chap. 2, the stability constraints are defined based on a known nonlinear state feedback controller. In this chapter, the constraint of the upper-tier networked LMPC design is based on the lower-tier output feedback controller. The upper-tier LMPC optimization problem is defined as follows:

$$\min_{u_a \in S(\Delta)} \int_{t_a}^{t_a + N\Delta} \left[\|\tilde{x}(\tau)\|_{Q_c} + \|u_s(\tau)\|_{R_{c1}} + \|u_a(\tau)\|_{R_{c2}} \right] d\tau, \tag{3.9}$$

$$\text{s.t.} \quad \dot{\tilde{x}}(t) = f\big(\tilde{x}(t), u_s(t), u_a(t), 0\big), \tag{3.10}$$

$$u_s(t) = k_s\big(h_s\big(\tilde{x}(t)\big)\big), \tag{3.11}$$

$$\dot{\hat{x}}(t) = f\big(\hat{x}(t), k_s\big(h_s\big(\hat{x}(t)\big)\big), 0, 0\big), \tag{3.12}$$

$$\tilde{x}(t_a) = \hat{x}(t_a) = x(t_a), \tag{3.13}$$

$$V\big(\tilde{x}(t)\big) \leq V\big(\hat{x}(t)\big), \quad \forall t \in [t_a, t_a + N\Delta), \tag{3.14}$$

where $x(t_a)$ is the state obtained from both $y_s(t_a)$ and $y_a(t_a)$, $\tilde{x}(t)$ is the predicted trajectory of the two-tier nominal system with u_a computed by this upper-tier LMPC, and $\hat{x}(t)$ is the predicted trajectory of the two-tier nominal system for the input trajectory $u_a(t) \equiv 0$ for all $t \in [t_a, t_a + N\Delta)$. The optimal solution to this optimization problem is denoted $u_a^*(t|t_a)$. This signal is defined for all $t \geq t_a$ with $u_a^*(t|t_a) = 0$ for all $t \geq t_a + N\Delta$.

The control inputs of the two-tier control architecture based on the above LMPC are defined as follows:

$$u_s(t) = k_s\big(h_s\big(x(t)\big)\big), \quad \forall t, \tag{3.15}$$

$$u_a(t) = u_a^*(t|t_a), \quad \forall t \in [t_a, t_{a+1}), \tag{3.16}$$

where $u_a^*(t|t_a)$ is the optimal solution of the LMPC of Eqs. 3.9–3.14 at time step t_a. This implementation technique takes into account that the lower-tier controller uses the continuously available measurements, while the upper-tier controller has to operate in open-loop between consecutive asynchronous measurements.

Note that the constraint of Eq. 3.14 in the LMPC of Eqs. 3.9–3.14 is needed to ensure that the value of the Lyapunov function of the closed-loop system under the two-tier control architecture is lower than or equal to the Lyapunov function of the closed-loop system when it is only controlled by the lower-tier controller. By imposing the constraint of Eq. 3.14, we can prove that the stability of the closed-loop system under the two-tier control architecture with inputs determined as in Eqs. 3.15–3.16 which is shown in Sect. 3.5.2.

Remark 3.9 By definition, $u_a^*(t|t_a) = 0$ for all $t \geq t_a + N\Delta$. This implies that the upper-tier controller switches off when it has been operating in open-loop for a large time, because in this case, the last received information is no longer useful to improve the performance of the lower-tier controller. The two-tier networked control architecture is (by design) stable because of the lower-tier controller stability properties. The main problem is how to improve the closed-loop performance using asynchronous communications in a way such that the stability properties of the closed-loop system under the lower-tier controller are not compromised. Setting the control input of the upper-tier controller to zero after a given time is necessary to maintain the stability properties, because after a sufficiently large time, the upper-tier input implemented in open-loop is not improving the closed-loop performance and may act as a disturbance.

3.5.2 Stability Properties

Combining the information from a hybrid communication system may lead to losing the stability properties of the lower-tier controller. The resulting closed-loop system is an asynchronous system [73] and we follow a Lyapunov-based approach to study the stability properties of the two-tier control architecture with the upper-tier controller design as in Eqs. 3.9–3.14. The main idea, is to compute the input $u_a(t)$ applied to the system in a way such that it is guaranteed that the value of the Lyapunov function at time instants t_a, $V(x(t_a))$, is a decreasing sequence of values with a lower bound. This guarantees practical stability of the closed-loop system. This property is presented in Theorem 3.1 below. To state this theorem, we need the following propositions.

Proposition 3.1 *Consider the system of Eqs. 3.1–3.3 in closed-loop with a lower-tier controller k_s. If k_s satisfies the conditions of Eqs. 3.6–3.8, there exists a $\mathcal{KL}$ function $\beta(r,s)$, a $\mathcal{K}$ function γ and a constant $\theta_{\max}$ such that if $x(t_0) \in \Omega_\rho$ and $u_a(t) = 0$ for all t then:*

$$V(x(t)) \le \beta(V(x(t_0)), t - t_0) + \gamma\left(\max_{\tau \in [t_0,t]} \|w(\tau)\|\right) \tag{3.17}$$

for all $w \in W$ with $\theta \le \theta_{\max}$.

This proposition provides us with a bound on the trajectories of the Lyapunov function of the state of the system of Eqs. 3.1–3.3 in closed-loop with the lower-tier controller and $u_a(t) = 0$. The proof of Proposition 3.1 can be found in [40].

Proposition 3.2 *Consider the following state trajectories:*

$$\dot{x}_a(t) = f(x_a(t), k_s(h_s(x_a(t))), u_a(t), w(t)), \tag{3.18}$$

$$\dot{x}_b(t) = f(x_b(t), k_s(h_s(x_b(t))), u_a(t), 0) \tag{3.19}$$

with initial states $x_a(t_0) = x_b(t_0) \in \Omega_\rho$. There exists a class $\mathcal{K}$ function f_W such that:

$$\|x_a(t) - x_b(t)\| \le f_W(t - t_0) \tag{3.20}$$

with:

$$f_W(\tau) = \frac{L_w \theta}{L'_x} \left(e^{L'_x \tau} - 1 \right) \tag{3.21}$$

for all $x_a(t), x_b(t) \in \Omega_\rho$ *and all* $w(t) \in W.$

Proof Define the error vector as $e(t) = x_a(t) - x_b(t)$. The time derivative of the error is given by:

$$\dot{e}(t) = f\left(x_a(t), k_s\left(h_s\left(x_a(t)\right)\right), u_a(t), w(t)\right)$$
$$- f\left(x_b(t), k_s\left(h_s\left(x_b(t)\right)\right), u_a(t), 0\right). \tag{3.22}$$

By the local Lipschitz property assumed for the vector field $f(x, u_s, u_a, w)$, there exist positive constants L_w, L_x and L_{u1} such that:

$$\left\| \dot{e}(t) \right\| \le L_w \left\| w(t) - 0 \right\| + L_x \left\| x_a(t) - x_b(t) \right\|$$
$$+ L_{u1} \left\| k_s\left(h_s\left(x_a(t)\right)\right) - k_s\left(h_s\left(x_b(t)\right)\right) \right\| \tag{3.23}$$

for all $x_a(t), x_b(t) \in \Omega_\rho$ and $w(t) \in W$. By continuity and smoothness properties of k_s and h_s, there exists a positive constant L_{u2} such that:

$$\left\| k_s\left(h_s\left(x_a(t)\right)\right) - k_s\left(h_s\left(x_b(t)\right)\right) \right\| \le L_{u2} \left\| x_a(t) - x_b(t) \right\| \tag{3.24}$$

for all $x_a(t), x_b(t) \in \Omega_\rho$. Thus the following inequality can be obtained from the inequality of Eq. 3.23:

$$\left\| \dot{e}(t) \right\| \le L_w \left\| w(t) \right\| + (L_x + L_{u1}L_{u2}) \left\| x_a(t) - x_b(t) \right\|$$
$$\le L_w \theta + (L_x + L_{u1}L_{u2}) \left\| e(t) \right\|. \tag{3.25}$$

Integrating $\left\| \dot{e}(t) \right\|$ with initial condition $e(t_0) = 0$ (recall that $x_a(t_0) = x_b(t_0)$), the following bound on the norm of the error vector is obtained:

$$\left\| e(t) \right\| \le \frac{L_w \theta}{L'_x} \left(e^{L'_x(t - t_0)} - 1 \right), \tag{3.26}$$

where $L'_x = L_x + L_{u1}L_{u2}$. This implies that the condition of Eq. 3.20 holds for:

$$f_W(\tau) = \frac{L_w \theta}{L'_x} \left(e^{L'_x \tau} - 1 \right), \tag{3.27}$$

which proves this proposition. $\square$

Theorem 3.1 *Consider the system of Eqs. 3.1–3.3 in closed-loop with* y_s *available for all* t, y_a *available at asynchronous time instants* $\{t_{a \ge 0}\}$ *without delay (i.e.,* $d_a \equiv 0$*) and a lower-tier controller* k_s *satisfying the conditions of Eqs. 3.6–3.8. Let the closed-loop system be controlled under the two-tier control architecture* *with the upper-tier LMPC of Eqs. 3.9–3.14 and control inputs determined as in Eqs. 3.15–3.16. If $x(t_0) \in \Omega_\rho$, $\theta \leq \theta_{max}$, $N \geq 1$, $\Delta > 0$ and there exist a concave function g such that:*

$$g(x) \geq \beta(x, N\Delta) \tag{3.28}$$

for all $x \in \Omega_\rho$, and a positive constant $c \leq \rho$ such that:

$$c - g(c) \geq f_V\big(f_W(N\Delta)\big) \tag{3.29}$$

with $f_V(\cdot)$ defined in Eq. 2.49 and $f_W(\cdot)$ defined in Eq. 3.21, then $x(t)$ is ultimately bounded in $\Omega_{\rho_c} \subseteq \Omega_\rho$ where:

$$\rho_c = \max\Big\{\max_c \beta(c, N\Delta) + f_V\big(f_W(N\Delta)\big), \gamma(\theta_{max})\Big\}. \tag{3.30}$$

Proof In order to prove that the closed-loop system is ultimately bounded in a region that contains the origin, we will prove that $V(x(t_a))$ is a decreasing sequence of values with a lower bound for the worst possible case, that is, the upper-tier controller always operates in open-loop for a period of time longer than $N\Delta$ between consecutive samples, that is, $t_{a+1} - t_a > N\Delta$ for all a. The trajectory $\hat{x}(t)$ corresponds to the nominal system in closed-loop with the lower-tier controller with initial state $x(t_a)$. Taking into account Proposition 3.1, the following inequality holds:

$$V\big(\hat{x}(t)\big) \leq \beta\big(V\big(x(t_a)\big), t - t_a\big). \tag{3.31}$$

The constraint of Eq. 3.14 of the upper-tier LMPC of Eqs. 3.9–3.14 guarantees that:

$$V\big(\tilde{x}(t)\big) \leq V\big(\hat{x}(t)\big), \quad \forall t \in [t_a, t_a + N\Delta). \tag{3.32}$$

Assuming that $x(t) \in \Omega_\rho$ for all times (which is automatically satisfied when the system is proved to be ultimately bounded below), we can apply Proposition 2.3 (presented in Chap. 2) to obtain the following inequalities:

$$V\big(x(t_a + N\Delta)\big) \leq V\big(\tilde{x}(t_a + N\Delta)\big) + f_V\big(\|x(t_a) - \tilde{x}(t_a)\|\big). \tag{3.33}$$

Applying Proposition 3.2, we obtain the following upper bound on the deviation of $\tilde{x}(t)$ from $x(t)$:

$$\big\|x(t_a + N\Delta) - \tilde{x}(t_a + N\Delta)\big\| \leq f_W(N\Delta). \tag{3.34}$$

Using the inequalities of Eqs. 3.31–3.34, the following upper bound on $V(x(t_a + N\Delta))$ is obtained:

$$V\big(x(t_a + N\Delta)\big) \leq \beta\big(V\big(x(t_a)\big), N\Delta\big) + f_V\big(f_W(N\Delta)\big). \tag{3.35}$$

Taking into account that for all $t \geq t_a + N\Delta$ the upper-tier controller is switched off, i.e., $u_a(t) = 0$, and only the lower-tier controller is in action, the following bound on $V(x(t_{a+1}))$ is obtained from Proposition 3.1:

$$V\big(x(t_{a+1})\big) \leq \max\big\{V\big(x(t_a + N\Delta)\big), \gamma(\theta_{max})\big\} \tag{3.36}$$

for all $w(t) \in W$. Because function $g(\cdot)$ is concave, $z - g(z)$ is an increasing function. If there is a positive constant $c \le \rho$ satisfying the condition of Eq. 3.29, then the condition of Eq. 3.29 holds for all $z > c$. Taking into account that $g(z) \ge \beta(z, N\Delta)$ for all $z \le \rho$, the following inequality is obtained:

$$z - \beta(z, N\Delta) \ge f_V\big(f_W(N\Delta)\big) \tag{3.37}$$

when $c \le z \le \rho$. From the inequality of Eq. 3.37 and the inequality of Eq. 3.35, we obtain that:

$$V\big(x(t_{a+1})\big) \le \max\big\{V\big(x(t_a)\big), \gamma(\theta_{max})\big\} \tag{3.38}$$

for all $V(x(t_k)) \ge c$. It follows using Lyapunov arguments that:

$$\limsup_{t \to \infty} V\big(x(t)\big) \le \rho_c, \tag{3.39}$$

where:

$$\rho_c = \max\big\{\max_c \beta(c, N\Delta) + f_V\big(f_W(N\Delta)\big), \gamma(\theta_{max})\big\}. \tag{3.40}$$

□

Remark 3.10 In general, the size of the region in which the state is ultimately bounded, depends on the prediction horizon $N\Delta$. The prediction horizon $N\Delta$ sets the maximum amount of time on which the upper-tier controller will be operating in open-loop.

Remark 3.11 Referring to Theorem 3.1, the assumption that there exists a concave function g such that $g(x) \ge \beta(x, N\Delta)$ imposes an upper bound on $N\Delta$ and is made, without any loss of generality, to simplify the proof of Theorem 3.1, that is, the result of Theorem 3.1 could still be proved without this assumption but the proof would be more involved. The assumption that there exists a positive constant $c \le \rho$ such that $c - g(c) \ge f_V(f_W(N\Delta))$ guarantees that the derivative of the Lyapunov function of the state of the closed-loop system outside the level set $V(x) = c$ is negative under the two-tier control architecture with the upper-tier LMPC of Eqs. 3.9–3.14.

Remark 3.12 As in all MPC schemes, it is not possible to provide quantitative results that guarantee that the performance of the closed-loop system is better than any other controller, unless an infinite horizon is used. It makes sense that the system in closed-loop with the two-tier control architecture has in general a better performance because the cost function is taken into account in the optimization problem of the upper-tier controller. The case studies in Sects. 3.5.3 and 3.5.4 provide results that demonstrate this point.

Remark 3.13 Note that in order to take advantage of the asynchronous measurements, an alternative to the two-tier control architecture is to control the system of

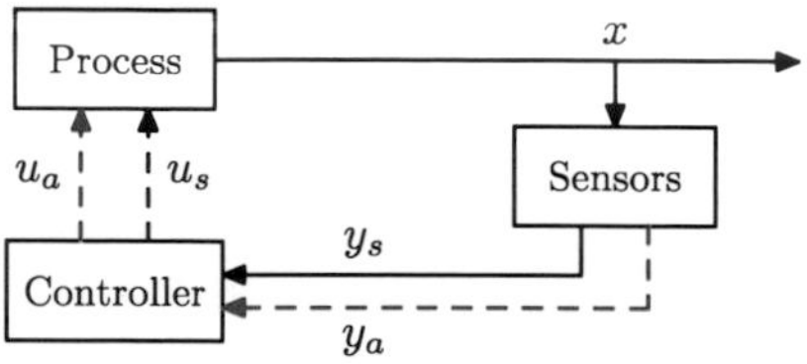

Fig. 3.3 Centralized
networked control system

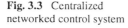

Eqs. 3.1–3.3 using a centralized MPC that calculate the input trajectories of both u_s and u_a at each asynchronous sampling time t_a when a new full state measurement is available by combining $y_s(t_a)$ and $y_a(t_a)$. Figure 3.3 shows a schematic of this kind of state feedback centralized control system. In particular, we may use the centralized LMPCs presented in Sects. 2.7 and 2.8 of Chap. 2 which are designed taking data losses or time-varying measurement delays explicitly into account, both in the optimization problem formulations and in the controller implementations. For the case that there is no time-varying delays in the asynchronous measurements (i.e., $y_a(t_a) = h_a(x(t_a))$ with $d_a = 0$), the centralized LMPC taking into account asynchronous measurements for the system of Eqs. 3.1–3.3 is based on the following optimization problem:

$$\min_{u_a, u_s \in S(\Delta)} \int_{t_a}^{t_a + N\Delta} \left[\|\tilde{x}(\tau)\|_{Q_c} + \|u_s(\tau)\|_{R_{c1}} + \|u_a(\tau)\|_{R_{c2}} \right] d\tau, \qquad (3.41)$$

$$\text{s.t.} \quad \dot{\tilde{x}}(t) = f\big(\tilde{x}(t), u_s(t), u_a(t), 0\big), \qquad (3.42)$$

$$\dot{\hat{x}}(t) = f\big(\hat{x}(t), k_s\big(h_s\big(\hat{x}(t_a + j\Delta)\big)\big), 0, 0\big),$$

$$\forall t \in \big[t_a + j\Delta, t_a + (j+1)\Delta\big), \qquad (3.43)$$

$$\tilde{x}(t) = \hat{x}(t) = x(t_k), \qquad (3.44)$$

$$V\big(\tilde{x}(t)\big) \leq V\big(\hat{x}(t)\big), \quad \forall t \in [t_a, t_a + N\Delta), \qquad (3.45)$$

where the lower-tier controller k_s is used to generate the reference trajectory $\hat{x}$ (k_s is implemented in a sample-and-hold fashion). The optimal solution to this optimization problem is denoted $u^*_{c,s}(t|t_a)$ and $u^*_{c,a}(t|t_a)$. These signals are defined for all $t \geq t_a$ with $u^*_{c,s}(t|t_a) = u^*_{c,s}(t_a + N\Delta|t_a)$ and $u^*_{c,a}(t|t_a) = u^*_{c,a}(t_a + N\Delta|t_a)$ for all $t \geq t_a + N\Delta$. The inputs of the closed-loop system of Eqs. 3.1–3.3 are defined as follows:

$$u_s(t) = u^*_{c,s}(t|t_a), \quad \forall t \in [t_a, t_{a+1}), \qquad (3.46)$$

$$u_a(t) = u^*_{c,a}(t|t_a), \quad \forall t \in [t_a, t_{a+1}). \qquad (3.47)$$

In Sects. 3.5.3 and 3.5.4, we denote this control design as the centralized LMPC.

3.5.3 Application to a Chemical Reactor

Consider the CSTR example described by Eqs. 2.62–2.63 introduced in Sect. 2.7.4. In this section, we consider a flow rate disturbance in the feed flow rate F of pure A, ΔF, and choose the rate of heat input or removal Q and the change of the inlet reactant A concentration ΔC_{A0} as the control inputs. The control objective is to stabilize the system at the open-loop unstable steady-state $T_s - 388$ K, $C_{As} - 3.59$ mol/l. The flow rate uncertainty is bounded by $|\Delta F| \leq 3$ m^3/h.

We assume that measurements of temperature T are available continuously, and the measurements of the concentration C_A are available asynchronously at time instants $\{t_{a \geq 0}\}$. We also assume that there exists a lower bound $\Delta_{\min}$ on the time interval between two consecutive concentration measurements.

In order to model the time sequence $\{t_{a \geq 0}\}$, we use a lower-bounded random Poisson process. The Poisson process is defined by the number of events per unit time W. The interval between two consecutive concentration sampling times (events of the Poisson process) is given by $\Delta_a = \max\{\Delta_{\min}, \frac{-\ln \chi}{W}\}$, where χ is a random variable with uniform probability distribution between 0 and 1. For the simulations carried out in this section we pick $\Delta_{\min} = 0.025$ h, which is meaningful from a practical point of view with respect to concentration measurements.

The CSTR model of Eqs. 2.62–2.63 belongs to the class of nonlinear systems described by the system of Eqs. 3.1–3.3 where $x^T = [x_1 \ x_2] = [T - T_s \ C_A - C_{As}]$ is the state, $u_s = Q$ and $u_a = \Delta C_{A0}$ are the manipulated inputs, $w = \Delta F$ is a time varying bounded disturbance, $y_s = x_1 = T - T_s$ is obtained from the continuous temperature measurement T and $y_a = x_2 = C_A - C_{As}$ is obtained from the asynchronously sampled concentration measurement C_A.

First, an output feedback controller (lower-tier controller) based on the continuous temperature measurements (i.e., x_1) is designed to stabilize the process using only the rate of heat input $u_s = Q$ as the manipulated input, which is bounded by $|u_s| \leq 10^5$ KJ/h. In particular, the following proportional-integral (PI) control law is used as the lower-tier controller:

$$u_s(t) = K\left(x_1(t) + \frac{1}{T_i} \int_0^t x_1(\tau)\,d\tau\right), \tag{3.48}$$

where K is the proportional gain and T_i is the integral time constant. To compute the parameters of the PI controller, the linearized model $\dot{x} = Ax + Bu_s$ of the CSTR of Eqs. 2.62–2.63 around the equilibrium point is obtained. The proportional gain K is chosen to be -8100. This value guarantees that the origin of $\dot{x} = (A + BK[1\ 0])x$ is asymptotically stable with its eigenvalues being $\lambda_1 = -1.06 \times 10^5$ and $\lambda_2 = -4.43$. A quadratic Lyapunov function $V(x) = x^T P x$ with:

$$P = \begin{bmatrix} 0.024 & 5.21 \\ 5.21 & 1.13 \times 10^3 \end{bmatrix} \tag{3.49}$$

Fig. 3.4 State and input
trajectories of the CSTR of
Eqs. 2.62–2.63 under the
lower-tier PI control of
Eq. 3.48

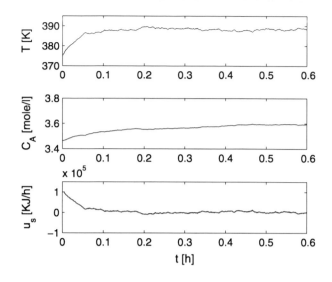

is obtained by solving an algebraic Lyapunov equation $A_c^T P + P A_c + Q_c = 0$ for
P with $A_c = A + BK[1\ 0]$ and Q_c being the following weighting matrix:

$$Q_c = \begin{bmatrix} 1 & 0 \\ 0 & 10^4 \end{bmatrix}. \tag{3.50}$$

This Lyapunov function will be used to design the upper-tier LMPC and the centralized LMPC. The integral time constant is chosen to be $T_i = 49.6$ h. For simplicity,
the Lyapunov function $V(x)$ is determined on the basis of the closed-loop system
under the proportional (P) term of the PI controller only; the effect of the integral
(I) term is very small for the specific choice of the controller parameters used in the
simulations. The state and input trajectories of the CSTR of Eqs. 2.62–2.63 starting
from $x_0 = [370\ 3.41]^T$ under the PI controller are shown in Fig. 3.4. From Fig. 3.4,
we see that the PI controller of Eq. 3.48 stabilizes the temperature and concentration of the CSTR of Eqs. 2.62–2.63 at the equilibrium point in about 0.1 h and 0.4 h,
respectively.

Next, we implemented the presented two-tier control architecture to improve the
performance of the closed-loop system. In this set of simulations, the PI controller
of Eq. 3.48 is used as the lower-tier controller. Instead of abandoning the less frequent concentration measurement, we take advantage of both the continuous measurements of the temperature T and the asynchronous concentration measurements
C_A together with the nominal model of the system of Eqs. 2.62–2.63 to design the
upper-tier LMPC of Eqs. 3.9–3.14. The inlet concentration change ΔC_{A0}, which is
bounded by $|\Delta C_{A0}| \leq 1$ kmol/m^3, is the manipulated input for the upper-tier LMPC.
In the design of the upper-tier LMPC, the performance index is defined by Q_c given
in Eq. 3.50 and $R_{c1} = R_{c2} = 0$. The values of the weights in Q_c have been chosen
to account for the different range of numerical values for each state. The sampling
time of the LMPC is $\Delta = 0.025$ h; the prediction horizon is $N = 11$ so that the
prediction captures most of the dynamic evolution of the process.

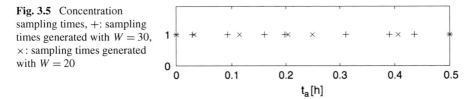

Fig. 3.5 Concentration sampling times, +: sampling times generated with $W = 30$, ×: sampling times generated with $W = 20$

The two-tier control architecture is implemented as discussed in Sect. 3.5.1. The lower-tier controller uses the continuous temperature measurements to control $u_s(t)$. When the measurements of T and C_A are obtained at time instant t_a, $x(t_a)$, is obtained from the two measurements. Based on the state $x(t_a)$, the LMPC optimization problem of Eqs. 3.9–3.14 is solved and an optimal input trajectory $u_a^*(t|t_a)$ is obtained. This optimal input trajectory is implemented until a new concentration measurement is obtained at time t_{a+1} (note that a indexes the number of concentration samples received, not a given sampling time). Note that because a PI controller is used in the lower-tier, we need to predict the controller dynamics (the control effects generated by the integral part) in the optimization problem of the LMPC.

The stability and robustness of the two-tier control architecture have been studied with two different initial conditions $x(0) = [370\ 3.41]^T$ and $x(0) = [375\ 3.46]^T$ associated with two different concentration measurement sequences $\{t_{a\geq0}\}$ (see Fig. 3.5) generated with $W = 30$ and $W = 20$, respectively. The average time intervals between two consecutive sampling times are 0.0625 h for $W = 30$ and 0.0833 h for $W = 20$. In addition, two different disturbance trajectories of $w(t)$ with a random value at each simulation step are added to the closed-loop system. The state and inputs trajectories of the CSTR of Eqs. 2.62–2.63 under the two-tier control architecture are shown in Fig. 3.6. From Fig. 3.6, we see that the two-tier control architecture stabilizes the temperature and concentration of the system in about 0.1 h and 0.05 h, respectively. This implies that the resulting closed-loop system response is faster compared with the speed of the closed-loop response under the PI controllers. Moreover, the cost associated with the resulting closed-loop trajectories is lower.

Another set of simulations was carried out to compare the two-tier control architecture with the lower-tier PI control system from a performance point of view. Table 3.1 shows the total cost computed for 20 different closed-loop simulations under the two-tier control architecture and the PI control. To carry out this comparison, we have computed the total cost of each simulation based on the performance index defined as follows:

$$\int_{t_0}^{t_f} \|x(\tau)\|_{Q_c}\, d\tau, \tag{3.51}$$

where $t_0 = 0$ is the initial time and $t_f = 0.5$ h is the length of the simulations. For this set of simulations W is chosen to be 10. For each pair of simulations (one for each control scheme) a different initial state inside the stability region, a different uncertainty trajectory and a different random concentration measurement sequence are chosen. As it can be seen in Table 3.1, the two-tier control architecture has a cost lower than the corresponding total cost under the PI controller in all the simulations.

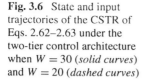

Fig. 3.6 State and input trajectories of the CSTR of Eqs. 2.62–2.63 under the two-tier control architecture when $W = 30$ (*solid curves*) and $W = 20$ (*dashed curves*)

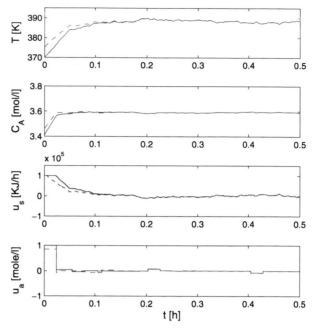

Table 3.1 Total performance costs along the closed-loop trajectories of the CSTR of Eqs. 2.62–2.63 under the local PI controller of Eq. 3.48 and the two-tier control with the upper-tier LMPC of Eqs. 3.9–3.14

sim.	Two-Tier	PI	sim.	Two-Tier	PI
1	203.92	704.54	11	224.03	831.63
2	188.74	815.47	12	203.78	738.47
3	198.33	922.87	13	265.44	617.15
4	221.76	640.87	14	210.58	704.95
5	240.44	656.47	15	190.68	723.05
6	226.44	847.43	16	209.66	695.60
7	199.19	779.03	17	205.90	808.71
8	233.40	736.65	18	211.29	749.24
9	200.45	702.26	19	214.79	737.62
10	198.74	753.25	20	217.13	813.70

We have also carried out another set of simulation to compare the presented two-tier scheme with a controller using the measurements of T and C_A to decide both control inputs u_s and u_a in the centralized LMPC of Eqs. 3.41–3.45; see Remark 3.13. This implies that this approach does not take full advantage of the continuous measurement of T. The LMPC of Eqs. 3.41–3.45 optimizes the future sampled input trajectory $u_a(t), u_s(t)$ with sampling time Δ. When at a time instant t_a, both the measurements of T and C_A are available (a state measurement is available), this

Fig. 3.7 State and input trajectories of the CSTR of Eqs. 2.62–2.63 under the centralized LMPC of Eqs. 3.41–3.45 with concentration sampling times generated with $W = 30$ (*solid curves*) and $W = 20$ (*dashed curves*)

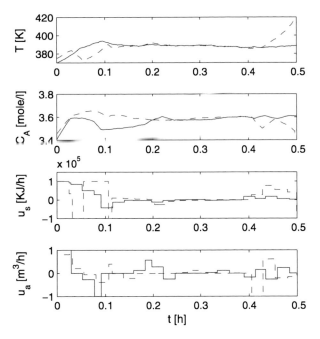

optimization problem is evaluated and two optimal input trajectories $u_{c,s}^*(t|t_a)$ and $u_{c,a}^*(t|t_a)$ are obtained and implemented until the next measurement of both T and C_A are available.

For this set of simulations, the centralized LMPC of Eqs. 3.41–3.45 uses the same parameters as the ones of the two-tier control architecture. The same initial conditions, concentration sampling times (see Fig. 3.5) and disturbance trajectories are used in this set of simulations. The state and inputs trajectories of the closed-loop system under the LMPC of Eqs. 3.41–3.45 are shown in Fig. 3.7. From Fig. 3.7, it can be seen that the centralized LMPC stabilizes the system (solid curves) when the time intervals between two consecutive measurements are small (0.0625 h), but loses stability and can not stabilize the system (dashed curves) when these time intervals get bigger (0.0833 h). The centralized LMPC of Eqs. 3.41–3.45 does not profit from the continuous measurements of the temperature, thus, the stability region of the closed-loop system is in general reduced to a much smaller one compared to that obtained under the two-tier control architecture.

Remark 3.14 The performance index considered in this example penalizes only the closed-loop system state and not the control action because the two-tier control architecture utilizes different manipulated inputs from the lower-tier PI controller and this would complicate the comparison if penalty on the control action is included in the cost. Since the performance index has only penalty on the closed-loop system state, we have included an input constraint on the upper-tier manipulated input, ΔC_{A0}, to avoid computation of unnecessarily large control actions by the upper-tier controller (i.e., $|u_a| \leq 1$ kmol/m^3).

Remark 3.15 Note that in this particular example, the improvement in the closed-loop performance is achieved due to the extra control input u_a which is guided by the LMPC of Eqs. 3.9–3.14 that uses all available measurements. Since PI controller is used as the lower-tier controller, the extra available asynchronous measurements would not have changed the closed-loop performance achieved by the lower-tier controller because the PI controller cannot use the extra measurements. This is also the case for the all the examples discussed in this chapter.

3.5.4 Application to a Reactor–Separator Process

Consider the reactor–separator process shown in Fig. 1.6 described in Sect. 1.2.3. Under the assumption that the three vessels have static holdup and other standard modeling assumptions, the dynamic equations describing the behavior of the system, obtained through material and energy balances, are given below [21]:

$$\frac{dx_{A1}}{dt} = \frac{F_{10}}{V_1}(x_{A10} - x_{A1}) + \frac{F_r}{V_1}(x_{Ar} - x_{A1}) - k_1 e^{\frac{-E_1}{RT_1}} x_{A1}, \tag{3.52}$$

$$\frac{dx_{B1}}{dt} = \frac{F_{10}}{V_1}(x_{B10} - x_{B1}) + \frac{F_r}{V_1}(x_{Br} - x_{B1}) + k_1 e^{\frac{-E_1}{RT_1}} x_{A1} - k_2 e^{\frac{-E_2}{RT_1}} x_{B1}, \tag{3.53}$$

$$\frac{dT_1}{dt} = \frac{F_{10}}{V_1}(T_{10} - T_1) + \frac{F_r}{V_1}(T_3 - T_1) + \frac{-\Delta H_1}{C_p} k_1 e^{\frac{-E_1}{RT_1}} x_{A1}$$
$$+ \frac{-\Delta H_2}{C_p} k_2 e^{\frac{-E_2}{RT_1}} x_{B1} + \frac{Q_1}{\rho C_p V_1}, \tag{3.54}$$

$$\frac{dx_{A2}}{dt} = \frac{F_1}{V_2}(x_{A1} - x_{A2}) + \frac{F_{20}}{V_2}(x_{A20} - x_{A2}) - k_1 e^{\frac{-E_1}{RT_2}} x_{A2}, \tag{3.55}$$

$$\frac{dx_{B2}}{dt} = \frac{F_1}{V_2}(x_{B1} - x_{B2}) + \frac{F_{20}}{V_2}(x_{B20} - x_{B2}) + k_1 e^{\frac{-E_1}{RT_2}} x_{A2} - k_2 e^{\frac{-E_2}{RT_2}} x_{B2}, \tag{3.56}$$

$$\frac{dT_2}{dt} = \frac{F_1}{V_2}(T_1 - T_2) + \frac{F_{20}}{V_2}(T_{20} - T_2) + \frac{-\Delta H_1}{C_p} k_1 e^{\frac{-E_1}{RT_2}} x_{A2}$$
$$+ \frac{-\Delta H_2}{C_p} k_2 e^{\frac{-E_2}{RT_2}} x_{B2} + \frac{Q_2}{\rho C_p V_2}, \tag{3.57}$$

$$\frac{dx_{A3}}{dt} = \frac{F_2}{V_3}(x_{A2} - x_{A3}) - \frac{F_r + F_p}{V_3}(x_{Ar} - x_{A3}), \tag{3.58}$$

$$\frac{dx_{B3}}{dt} = \frac{F_2}{V_3}(x_{B2} - x_{B3}) - \frac{F_r + F_p}{V_3}(x_{Br} - x_{B3}), \tag{3.59}$$

$$\frac{dT_3}{dt} = \frac{F_2}{V_3}(T_2 - T_3) + \frac{Q_3}{\rho C_p V_3}. \tag{3.60}$$

Table 3.2 Process variables of the reactor–separator process of Eqs. 3.52–3.63

x_{A1}, x_{A2}, x_{A3}	Mass fractions of A in vessels 1, 2, 3
x_{B1}, x_{B2}, x_{B3}	Mass fractions of B in vessels 1, 2, 3
x_{C1}	Mass fraction of C in vessel 3
x_{Ar}, x_{Br}, x_{Cr}	Mass fractions of A, B, C in the recycle
T_1, T_2, T_3	Temperatures in vessels 1, 2, 3
T_{10}, T_{20}	Feed stream temperatures to vessels 1, 2
F_1, F_2	Effluent flow rate from vessels 1, 2
F_{10}, F_{20}	Feed stream flow rates to vessels 1, 2
F_r, F_p	Flow rates of the recycle and purge
V_1, V_2, V_3	Volumes of vessels 1, 2, 3
E_1, E_2	Activation energy for reactions 1, 2
k_1, k_2	Pre-exponential values for reactions 1, 2
$\Delta H_1, \Delta H_2$	Heats of reaction for reactions 1, 2
$\alpha_A, \alpha_B, \alpha_C$	Relative volatilities of A, B, C
Q_1, Q_2, Q_3	Heat inputs into vessels 1, 2, 3
C_p, R, ρ	Heat capacity, gas constant and solution density

The model of the flash tank separator was derived under the assumption that the relative volatility for each of the species remains constant within the operating temperature range of the flash tank. This assumption allows calculating the mass fractions in the overhead based upon the mass fractions in the liquid portion of the vessel. It has also been assumed that there is a negligible amount of reaction taking place in the separator. The following algebraic equations model the composition of the overhead stream relative to the composition of the liquid holdup in the flash tank:

$$x_{Ar} = \frac{\alpha_A x_{A3}}{\alpha_A x_{A3} + \alpha_B x_{B3} + \alpha_C x_{C3}}, \tag{3.61}$$

$$x_{Br} = \frac{\alpha_B x_{B3}}{\alpha_A x_{A3} + \alpha_B x_{B3} + \alpha_C x_{C3}}, \tag{3.62}$$

$$x_{Cr} = \frac{\alpha_C x_{C3}}{\alpha_A x_{A3} + \alpha_B x_{B3} + \alpha_C x_{C3}}. \tag{3.63}$$

The definitions for the variables used in Eqs. 3.52–3.63 and the corresponding parameter values used in this example can be found in Tables 3.2 and 3.3, respectively. Note that the reactions $A \rightarrow B$ and $B \rightarrow C$ are referred to as reactions 1 and 2, respectively.

Each of the tanks in the process has an external heat input. The manipulated inputs to the system are the heat inputs to the three vessels, Q_1, Q_2 and Q_3, and the feed stream flow rate to vessel 2, F_{20}.

We assume that the measurements of temperatures T_1, T_2 and T_3 are available continuously, and the measurements of mass fractions $x_{A1}, x_{B1}, x_{A2}, x_{B2}, x_{A3}$ and x_{B3} are available asynchronously at time instants $\{t_{a\geq0}\}$. The same method used in

Table 3.3 Process parameters of the reactor–separator process of Eqs. 3.52–3.63

T_{10}	300 [K]	k_1	2.77×10^3 [s^{-1}]
T_{20}	300 [K]	k_2	2.5×10^3 [s^{-1}]
F_{10}	5.04 [m^3/h]	ΔH_1	-6×10^4 [KJ/kmol]
F_r	50.4 [m^3/h]	ΔH_2	-7×10^4 [KJ/kmol]
F_p	5.04 [m^3/h]	α_A	3.5
V_1	1.0 [m^3]	α_B	1
V_2	0.5 [m^3]	α_C	0.5
V_3	1.0 [m^3]	C_p	4.2 [KJ/kg K]
E_1	5×10^4 [KJ/kmol]	R	8.314 [KJ/kmol K]
E_2	6×10^4 [KJ/kmol]	ρ	1000 [kg/m^3]

Table 3.4 Steady-state operation parameters of x_{s1} and x_{s2} of the reactor–separator process of Eqs. 3.52–3.63

x_{s1}		x_{s2}	
Q_{1s}	12.6×10^5 [KJ/h]	Q_{1s}	12.6×10^5 [KJ/h]
Q_{2s}	16.2×10^5 [KJ/h]	Q_{2s}	13.32×10^5 [KJ/h]
Q_{3s}	12.6×10^5 [KJ/h]	Q_{3s}	11.88×10^5 [KJ/h]
F_{20s}	5.04 [m^3/h]	F_{20s}	5.04 [m^3/h]

Table 3.5 Steady-states x_{s1} and x_{s2} of the reactor–separator process of Eqs. 3.52–3.63

	x_{A1s}	x_{B1s}	T_{1s}	x_{A2s}	x_{B2s}	T_{2s}	x_{A3s}	x_{B3s}	T_{3s}
x_{s1}	0.383	0.581	447.8	0.391	0.572	444.6	0.172	0.748	449.6
x_{s2}	0.605	0.386	425.9	0.605	0.386	422.6	0.346	0.630	427.3

the example in Sect. 3.5.3 is used in this example to generate the time sequence $\{t_{a \geq 0}\}$.

For each set of steady-state inputs Q_{1s}, Q_{2s}, Q_{3s} and F_{20s} corresponding to a different operation condition, the system of Eqs. 3.52–3.63 has one stable steady-state x_s^T. In this example, we will study two different operating conditions corresponding to two different steady-states x_{s1} and x_{s2}. The parameters of the steady-state operation points and the values of the two steady-states are given in Table 3.4 and Table 3.5. The control objective is to steer the system to the steady-states from the initial state:

$$x(0)^T = [0.890, 0.110, 388.732, 0.886, 0.113, 386.318, 0.748, 0.251, 390.570]. \tag{3.64}$$

The system of Eqs. 3.52–3.63 belongs to the class of nonlinear systems described by the system of Eqs. 3.1–3.3 where $x^T = [x_1\ x_2\ x_3\ x_4\ x_5\ x_6\ x_7\ x_8\ x_9] = [x_{A1} - x_{A1s}\ x_{B1} - x_{B1s}\ T_1 - T_{1s}\ x_{A2} - x_{A2s}\ x_{B2} - x_{B2s}\ T_2 - T_{2s}\ x_{A3} - x_{A3s}\ x_{B3} - x_{B3s}\ T_3 - T_{3s}]$ is the state, $u_s^T = [u_{s1}\ u_{s2}\ u_{s3}] = [Q_1 - Q_{1s}\ Q_2 - Q_{2s}\ Q_3 - Q_{3s}]$ and $u_a = F_{20} - F_{20s}$ are the manipulated inputs, $y_s^T = [y_{s1}\ y_{s2}\ y_{s3}] = [x_3\ x_6\ x_9]$ is obtained

Table 3.6 Control parameters for steady-states x_{s1} and x_{s2} of the reactor–separator process of Eqs. 3.52–3.63	x_{s1}		x_{s2}	
	K_1	-5000	K_1	-5000
	K_2	-5000	K_2	-5000
	K_3	-5000	K_3	-5000
	T_i	5 [h]	T_i	5 [h]

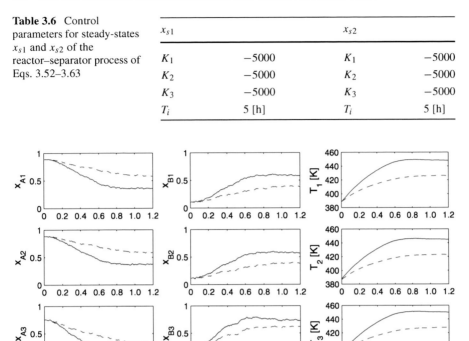

Fig. 3.8 State trajectories of the reactor–separator process of Eqs. 3.52–3.63 under lower-tier control law for steady-state x_{s1} (*solid curves*) and steady-state x_{s2} (*dashed curves*)

from the continuous temperature measurements and $y_a^T = [x_1\ x_2\ x_4\ x_5\ x_7\ x_8]$ is obtained from the asynchronously sampled mass fraction measurement. Time varying bounded process noise was added to the simulations.

Based on the continuous temperature measurements (i.e., y_s), three PI controllers (lower-tier controllers) are first designed following the Eq. 3.48 to stabilize the system of Eqs. 3.52–3.63 from the initial state $x(0)$ to the steady-state x_s using only the heat inputs as the manipulated inputs, which are bounded by $|Q_i| \le 2 \times 10^6$ KJ/h ($i = 1, 2, 3$). Using the same method as described in Sect. 3.5.3, the parameters of the PI controllers are obtained as shown in Table 3.6; and two different quadratic Lyapunov functions are obtained, one for each steady state x_{s1}, x_{s2}. The two Lyapunov functions are used to design the upper-tier LMPC controller and the centralized LMPC of Eqs. 3.41–3.45. The state and input trajectories of the system of Eqs. 3.52–3.63 under the lower-tier PI control are shown in Figs. 3.8 and 3.9. From Fig. 3.8, we see that the PI control law stabilizes the temperatures and mass fractions in the three vessels in about 0.7 h for both steady-states.

We design next the upper-tier LMPC of Eqs. 3.9–3.14 and the corresponding two-tier control architecture. The feed flow rate to vessel 2, $u_a = F_{20} - F_{20s}$, is the manipulated input for the upper-tier LMPC, which is bounded by $1 \le F_{20} \le 9$ m^3/h.

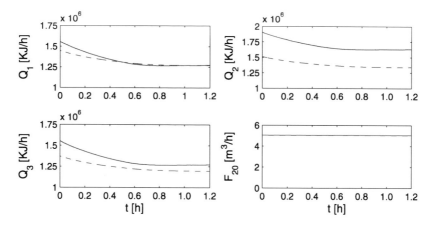

Fig. 3.9 Inputs trajectories of the reactor–separator process of Eqs. 3.52–3.63 under lower-tier control law for steady-state x_{s1} (*solid curves*) and steady-state x_{s2} (*dashed curves*)

Fig. 3.10 Mass fractions sampling times generated with $W = 1$ (+) and $W = 0.5$ (×)

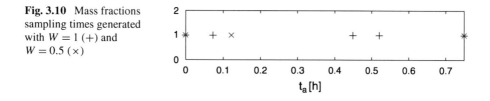

The performance index is defined by Q_c being the following weighting matrix:

$$Q_c = diag\left(\begin{bmatrix} 10^4 & 10^4 & 1 & 10^4 & 10^4 & 1 & 10^4 & 10^4 & 1 \end{bmatrix}\right) \qquad (3.65)$$

and $R_{c1} = R_{c2} = 0$. The sampling time of the LMPC is $\Delta = 0.025$ h and the prediction horizon is $N = 15$.

Two different simulations have been carried out with different mass fraction measurement sequences $\{t_{a\geq0}\}$ (see Fig. 3.10) generated with $W = 1$ and $W = 0.5$ for steady-states x_{s1} and x_{s2}, respectively. The average time intervals between two consecutive sampling times are 0.188 h for $W = 1$ and 0.375 h for $W = 0.5$. The state and input trajectories of the reactor–separator process of Eqs. 3.52–3.63 under the two-tier control architecture are shown in Figs. 3.11 and 3.12. Figure 3.11 shows that the two-tier control architecture stabilizes the temperatures and the mass fractions of the system in about 0.3 h. This implies that the resulting closed-loop system response is faster relative to the speed of the closed-loop response under the low-tier PI controllers.

Another set of simulations was also carried out to compare the two-tier control architecture with the lower-tier controller from a performance point of view. Table 3.7 shows the total cost computed for 10 different closed-loop simulations under the two-tier control architecture and the lower-tier controller. To carry out this comparison, we have computed the total cost of each simulation based on the performance index defined in Eq. 3.51 with Q_c given in Eq. 3.65 with different operation condi-

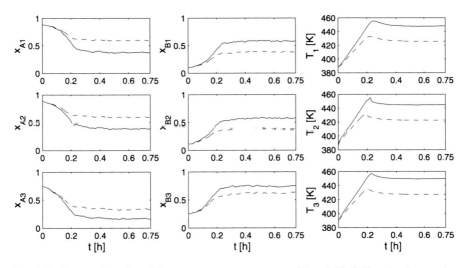

Fig. 3.11 State trajectories of the reactor–separator process of Eqs. 3.52–3.63 under the two-tier control architecture when $W = 1$ (*solid curves*) and $W = 0.5$ (*dashed curves*)

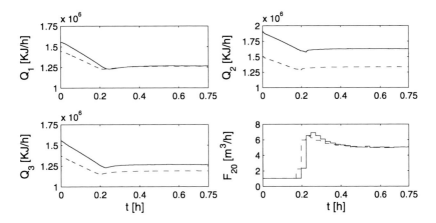

Fig. 3.12 Inputs trajectories of the reactor–separator process of Eqs. 3.52–3.63 under the two-tier control architecture when $W = 1$ (*solid curves*) and $W = 0.5$ (*dashed curves*)

tions in a simulation length of $t_f = 0.75$ h. For this set of simulations W is 1. As it can be seen in Table 3.7, the two-tier control architecture has a cost lower than the corresponding total cost under the lower-tier controller in all the simulations.

We have also carried out another set of simulations to compare the computational time needed to evaluate the upper-tier LMPC of Eqs. 3.9–3.14 with that of the centralized LMPC of Eqs. 3.41–3.45. For these simulations, the centralized LMPC uses the same parameters as the ones of the upper-tier LMPC in the present example. The simulations have been carried out using MATLAB® in a PENTIUM® 3.20 GHz. The nonlinear optimization problem has been solved using the function *fmincom*. To integrate the system model of Eqs. 3.52–3.63, both in the simulations and in the

Table 3.7 Total performance costs along the closed-loop trajectories of the reactor–separator process of Eqs. 3.52–3.63 under the local PI controller and the two-tier control with the upper-tier LMPC of Eqs. 3.9–3.14

sim.	Two-Tier	PI	sim.	Two-Tier	PI
1	1.179×10^5	2.760×10^5	6	1.560×10^5	3.742×10^5
2	1.164×10^5	2.795×10^5	7	1.645×10^5	3.951×10^5
3	1.273×10^5	2.991×10^5	8	1.701×10^5	4.107×10^5
4	1.351×10^5	3.177×10^5	9	1.962×10^5	4.408×10^5
5	1.364×10^5	3.240×10^5	10	1.848×10^5	4.492×10^5

optimization algorithm, an Euler method with a fixed integration time of 0.001 h has been implemented in C programming language. The mean time to solve the LMPC optimization problem of this set of simulations is 23.24 s for the upper-tier LMPC and 37.59 s for the centralized LMPC. From this set of simulations, we see that the computational time needed to solve the centralized LMPC optimization problem is substantially larger even though the closed-loop performance in terms of the total performance cost is comparable to the one of the two-tier control architecture. This is because the centralized LMPC has to optimize both the inputs u_s and u_a.

3.6 Two-Tier Networked Control Architecture with Continuous/Delayed Measurements

In this section, we extend the design of two-tier networked control architecture presented in the previous section for the system of Eqs. 3.1–3.3 with continuous and asynchronous measurements involving time-varying delays (i.e., $d_a \neq 0$).

3.6.1 Upper-Tier Networked LMPC Formulation

At each time instant t_a when a new asynchronous measurement $y_a(t_a - d_a)$ is received, a delayed state measurement $x(t_a - d_a)$ is obtained by combining this measurement with the previously received synchronous measurement $y_s(t_a - d_a)$. Based on this delayed state measurement $x(t_a - d_a)$, the nominal model of the system of Eqs. 3.1–3.3, the continuous measurements $y_s(t)$ and the control inputs applied from $t_a - d_a$ to t_a, an estimate of the current state $\tilde{x}(t_a)$ is computed. Note that this implies that the upper-tier controller has to store its past control input trajectory, know the explicit expression and parameters of the lower-tier controller and use the continuous measurements $y_s(t)$ to predict the control inputs carried out by the lower-tier controller. The estimated state $\tilde{x}(t_a)$ is then used to obtain the optimal future control input trajectory of u_a by means of an LMPC optimization problem. This input trajectory is implemented until a new measurement arrives at time t_{a+1}. If the time between two consecutive measurements is longer than the prediction horizon, u_a is set to zero until a new measurement arrives and the optimal control problem is solved again. Specifically, the upper-tier LMPC optimization problem taking into account delays in asynchronous measurements is defined as follows:

$$\min_{u_a \in S(\Delta)} \int_{t_a}^{t_a+N\Delta} \left[\|\tilde{x}(\tau)\|_{Q_c} + \|u_s(\tau)\|_{R_{c1}} + \|u_a(\tau)\|_{R_{c2}} \right] d\tau, \tag{3.66}$$

$$\dot{\tilde{x}}(t) = f\left(\tilde{x}(t), u_s(t), u_a(t), 0\right), \quad \forall t \in [t_a - d_a, t_a + N\Delta), \tag{3.67}$$

$$u_s(t) = k_s\left(h_s\left(\tilde{x}(t)\right)\right), \tag{3.68}$$

$$u_a(t) = u_a^*(t), \quad \forall t \in [t_a - d_a, t_a), \tag{3.69}$$

$$\tilde{x}(t_a - d_a) = x(t_a - d_a), \tag{3.70}$$

$$\dot{\hat{x}}(t) = f\left(\hat{x}(t), k_s\left(h_s\left(\hat{x}(t)\right)\right), 0, 0\right), \quad t \in [t_a, t_a + N\Delta), \tag{3.71}$$

$$\hat{x}(t_a) = \tilde{x}(t_a), \tag{3.72}$$

$$V\left(\tilde{x}(t)\right) \le V\left(\hat{x}(t)\right), \quad \forall t \in [t_a, t_a + N\Delta), \tag{3.73}$$

where $u_a^*(t)$ indicates the actual input trajectory of u_a that has been applied to the system, $x(t_a - d_a)$ is the state obtained combining both the measurements of $y_s(t_a - d_a)$ and $y_a(t_a - d_a)$, and $\tilde{x}(t_a)$ is an estimate of the current system state. The optimal solution to this optimization problem is denoted $u_d^*(t|t_a)$. This signal is defined for all $t \ge t_a$ with $u_d^*(t|t_a) = 0$ for all $t \ge t_a + N\Delta$.

The control inputs of the two-tier control architecture based on the above LMPC are defined as follows:

$$u_s(t) = k_s\left(h_s\left(x(t)\right)\right), \quad \forall t, \tag{3.74}$$

$$u_a(t) = u_d^*(t|t_a), \quad \forall t \in [t_a, t_{a+1}), \tag{3.75}$$

where $u_d^*(t|t_a)$ is the optimal solution of the LMPC of Eqs. 3.66–3.73 at time step t_a.

Remark 3.16 In the LMPC of Eqs. 3.66–3.73 both the estimation of $x(t_a)$ from $x(t_a - d_a)$ and the evaluation of the future optimal input trajectory in $[t_a, t_{a+1})$ are carried out at the same time. First, the constraints of the problem guarantee that $\tilde{x}(t_a)$ has been estimated using the nominal model (the constraint of Eq. 3.67) and the actual inputs applied to the system (the constraint of Eq. 3.69) from the initial state $x(t_a - d_a)$ (the constraint of Eq. 3.70). Once the current state is estimated, the future input trajectory is optimized to minimize the cost function taking into account the actions of the lower-tier controller (the constraint of Eq. 3.71) while guaranteeing that a Lyapunov-based constraint is satisfied (the constraint of Eq. 3.73). The optimization problem of Eqs. 3.66–3.73 has been presented in order to get a compact controller formulation. It is possible to decouple the observer and the LMPC optimization problem as long as the observer provides an upper bound on the estimation error of $x(t_a)$. For example, a high-gain observer can be used to estimate

Fig. 3.13 Possible worst
scenario of the delayed
measurements received by the
networked controller and the
corresponding state
trajectories defined in the
LMPC of Eqs. 3.66–3.73

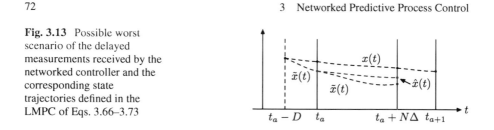

$x(t_a)$ from the continuous measurements and the applied inputs, and then use this estimated state to define the LMPC optimization problem.

Remark 3.17 The constraints of Eqs. 3.67 and 3.73 are a key element of the two-tier control architecture. In general, guaranteeing closed-loop stability of a decentralized control system is a difficult task because of the interactions between the different controllers and can only be done under certain assumptions (see, for example, [8, 92]). The constraint of Eq. 3.67 guarantees that the upper-tier controller takes into account the effect of the lower-tier controller to the applied inputs (recall that the lower-tier controller is designed without taking u_a into account). The constraint of Eq. 3.73 is used to guarantee that the value of the Lyapunov function is a decreasing sequence of time with a lower bound.

3.6.2 Stability Properties

In this subsection, we prove the stability result of the two-tier control architecture with the upper-tier LMPC of Eqs. 3.66–3.73.

Theorem 3.2 *Consider the system of Eqs. 3.1–3.3 in closed-loop with y_s available for all t, y_a available at asynchronous time instants $\{t_{a\geq 0}\}$ involving time-varying delays such that $d_a \leq D$ for all $a \geq 0$ and a lower-tier controller k_s satisfying the conditions of Eqs. 3.6–3.8. Let the closed-loop system be controlled under the two-tier control architecture with the upper-tier LMPC of Eqs. 3.66–3.73 and control inputs determined as in Eqs. 3.74–3.75. If $x(t_0) \in \Omega_\rho$, $\theta \leq \theta_{max}$, $N \geq 1$, $\Delta > 0$ and there exist a concave function g such that:*

$$g(x) \geq \beta\big(x + f_V\big(f_W(D)\big), N\Delta\big) \tag{3.76}$$

for all $x \in \Omega_\rho$, and a positive constant $c \leq \rho$ such that:

$$c - g(c) \geq f_V\big(f_W(D + N\Delta)\big) \tag{3.77}$$

with $f_V(\cdot)$ defined in Eq. 2.49 and $f_W(\cdot)$ defined in Eq. 3.21, then $x(t)$ is ultimately bounded in $\Omega_{\rho_d} \subseteq \Omega_\rho$ where:

$$\rho_d = \max\big\{\max_c \beta\big(c + f_V\big(f_W(D)\big), N\Delta\big) + f_V\big(f_W(D + N\Delta)\big), \gamma(\theta_{max})\big\}. \tag{3.78}$$

Proof In order to prove that the system of Eqs. 3.1–3.3 in closed-loop under the two-tier control architecture with the upper-tier LMPC of Eqs. 3.66–3.73 is ultimately bounded in a region that contains the origin, we will prove that the value of the Lyapunov function at times $\{t_{a \geq 0}\}$, $V(x)$, is a decreasing sequence of values with a lower bound on its magnitude for the worst possible case from a communication point of view, and hence for all possible sequences of measurement times and delays. The worst possible case from the communications point of view is that the measurements used to evaluate the upper tier LMPC are always received with the maximum delay D; that is $d_a = D$ for all a, and that the upper-tier LMPC always operates in open-loop for a period of time longer than $N\Delta$ between consecutive sampling times, that is, $t_{a+1} - t_a > N\Delta$ for all a. If the measurements are received with a smaller delay or more often, the LMPC has more precise information of the state of the system.

Figure 3.13 shows the worst case scenario for a system of dimension 1. Solid vertical lines are used to indicate the times at which new measurements are obtained (t_a and t_{a+1}) and when the upper-tier controller switches off at time $t_a + N\Delta$. The dashed vertical line indicates the time corresponding to the measurement obtained at t_a (that is, $t_a - D$). In this figure, three different state trajectories are shown. The actual state trajectory of the system of Eqs. 3.1–3.3 (including the uncertainty) is denoted as $x(t)$. The estimated state trajectory from $t_a - D$ to t_a and the predicted sampled trajectory under the two-tier control architecture with the upper-tier LMPC of Eqs. 3.66–3.73 along the prediction horizon with initial state the estimated state are denoted as $\tilde{x}(t)$. The nominal trajectory under the lower-tier controller k_s with $u_a \equiv 0$ along the prediction horizon with initial state the estimated state $\tilde{x}(t_a)$ is denoted as $\hat{x}(t)$. The state trajectories $\tilde{x}(t)$ and $\hat{x}(t)$ are obtained using the nominal model as defined in the LMPC optimization problem of Eqs. 3.66–3.73.

The trajectory $\hat{x}(t)$ corresponds to the nominal system in closed-loop with the lower-tier controller with initial state $\tilde{x}(t_a)$. Taking into account Proposition 3.1 the following inequality holds:

$$V\big(\hat{x}(t)\big) \leq \beta\big(V\big(\tilde{x}(t_a)\big), t - t_a\big). \tag{3.79}$$

The constraint of Eq. 3.73 guarantees that:

$$V\big(\tilde{x}(t)\big) \leq V\big(\hat{x}(t)\big), \quad \forall t \in [t_a, t_a + N\Delta). \tag{3.80}$$

Taking into account the constraints of Eqs. 3.67 and 3.70 and that the closed-loop trajectories are defined by the following equation:

$$\dot{x}(t) = f\big(x(t), k_s\big(h_s\big(x(t)\big)\big), u_a(t), w(t)\big), \tag{3.81}$$

we can apply Proposition 3.2 to obtain the following upper bounds on the deviation of $\tilde{x}(t)$ from $x(t)$:

$$\big\|x(t_a) - \tilde{x}(t_a)\big\| \leq f_W(D), \tag{3.82}$$

$$\big\|x(t_a + N\Delta) - \tilde{x}(t_a + N\Delta)\big\| \leq f_W(\tau_f + D). \tag{3.83}$$

Note that in Eqs. 3.82–3.83, Proposition 3.2 is used to obtain a bound on the difference between $\tilde{x}$ and x from $t_a - d_a$ to t_a to simplify the notation and the proof. Note that from $t_a - d_a$ to t_a, the real trajectory of u_s is applied to evaluate $\tilde{x}$, so a tighter bound on the difference between $\tilde{x}$ and x can be obtained. As mentioned before, the estimation of $x(t_a)$ can be done using any observer which provides a bound on the estimation error.

From Proposition 2.3 and the above inequalities, we obtain the following inequalities:

$$V\big(\tilde{x}(t_a)\big) \leq V\big(x(t_a)\big) + f_V\big(f_W(D)\big), \tag{3.84}$$

$$V\big(x(t_a + N\varDelta)\big) \leq V\big(\tilde{x}(t_a + N\varDelta)\big) + f_V\big(f_W(D + N\varDelta)\big). \tag{3.85}$$

From the inequalities of Eqs. 3.79–3.85, the following upper bound on $V(x(t_a + N\varDelta))$ is obtained:

$$V\big(x(t_a + N\varDelta)\big) \leq \beta\big(V\big(x(t_a)\big) + f_V\big(f_W(D)\big), N\varDelta\big) + f_V\big(f_W(D + N\varDelta)\big). \tag{3.86}$$

Taking into account that for all $t > t_a + N\varDelta$ the upper-tier controller is switched off, i.e., $u_a(t) = 0$, and only the lower-tier controller is in action, the following bound on $V(x(t_{a+1}))$ is obtained from Proposition 3.1:

$$V\big(x(t_{a+1})\big) \leq \max\big\{V\big(x(t_a + N\varDelta)\big), \gamma(\theta_{\max})\big\} \tag{3.87}$$

for all $w(t) \in W$. Because function $g(\cdot)$ is concave, $z - g(z)$ is an increasing function. If there is a constant $c_0 \leq c \leq \rho$ satisfying the condition of Eq. 3.77, then the condition of Eq. 3.77 holds for all $z > c$. Taking into account that $g(z) \geq \beta(z + f_V(f_W(D)), N\varDelta)$ for all $z \leq \rho$, the following inequality is obtained:

$$z - \beta\big(z + f_V\big(f_W(D)\big), N\varDelta\big) \geq f_V\big(f_W(D + N\varDelta)\big) \tag{3.88}$$

when $c \leq z \leq \rho$. From this inequality and the inequality of Eq. 3.87, we obtain that:

$$V\big(x(t_{a+1})\big) \leq \max\big\{V\big(x(t_a)\big), \gamma(\theta_{\max})\big\} \tag{3.89}$$

for all $V(x(t_a)) \geq c$. It follows using Lyapunov arguments that:

$$\limsup_{t \to \infty} V\big(x(t)\big) \leq \rho_d, \tag{3.90}$$

where:

$$\rho_d = \max\big\{\max_c \beta\big(c + f_V\big(f_W(D)\big), N\varDelta\big) + f_V\big(f_W(D + N\varDelta)\big), \gamma(\theta_{\max})\big\}. \tag{3.91}$$

□

3.6.3 Application to a Chemical Reactor

Consider the CSTR of Eqs. 2.62–2.63 discussed in Sects. 2.7.4 and 3.5.3. In the current section, we assume that $y_s = x_1 = T - T_s$ is obtained from the continuous temperature measurements T, and $y_a = x_2 = C_A - C_{As}$ is obtained at time instants $\{t_{a \geq 0}\}$ from the asynchronously sampled concentration measurement C_A subject to time-varying measurement delays. We also have a lower bound $T_{min} = 0.15$ h on the time interval between two consecutive concentration measurements and an upper bound D on the size of the delay; both will be computed via simulations even though conservative estimates could be computed from the theoretical results.

We use a lower-bounded Poisson process to model the time sequence $\{t_{a \geq 0}\}$ as discussed in Sect. 3.5.3. In order to model the delay size sequence $\{d_{a \geq 0}\}$, the size of delay associated with the concentration measurement at t_a is modeled by an upper-bounded random process given by $d_a = \min\{D, \phi H\}$, where ϕ is a uniformly distributed variable between 0 and 1, and $H = t_a - t_{a-1} + d_{a-1}$ is the size of the time interval between current time t_a and the time corresponding to the last concentration measurement $t_{a-1} - d_{a-1}$. This generation method guarantees that $d_a \leq D$ for all a. We assume that the initial state is known; that is, $d_0 = 0$ and $t_0 = 0$.

We use the lower-tier PI controller of Eq. 3.48 which is based on the continuous temperature measurements, and the same Lyapunov function $V(x) = x^T P x$. We implemented the two-tier control architecture with the LMPC of Eqs. 3.66–3.73 to improve the performance of the closed-loop system obtained under PI-only control. For the simulations carried out in this subsection, we pick the delay of each measurement to be $d_a = D = 0.15$ h for all a. These settings correspond to the worst-case effect from a communication point of view. For the other simulation settings, we use the ones used in Sect. 3.5.3 except that the prediction horizon is chosen to be $N = 6$. Note that the minimum time interval between two consecutive concentration measurements T_{min} is fixed by the system dynamics and the prediction horizon is set be equal to the minimum time interval between two consecutive y_a measurements, that is $N\Delta = T_{min}$.

The two-tier control architecture is implemented as discussed in the previous section. The lower-tier controller uses the continuous temperature measurements to decide $u_s(t)$. When a new measurement of C_A is obtained at time instant t_a with delay D, an estimate of the state of the CSTR, $x(t_a - D)$, is obtained by combining the concentration measurement and the previously received continuous measurement of the temperature T. Based on the state $x(t_a - D)$, the model of the process and the control actions applied, an estimate of the current state $\tilde{x}(t_a)$ is obtained. Based on this state estimate $\tilde{x}(t_a)$, the LMPC of Eqs. 3.66–3.73 is solved and an optimal input trajectory of u_a is obtained. This optimal input trajectory is implemented until a new concentration measurement is obtained at time t_{a+1}.

A simulation of the closed-loop system under the two-tier control architecture with the same initial condition $x(0) = [370 \; 3.41]^T$ has been carried out. The sampling sequence $\{t_{a \geq 0}\}$ generated with $W = 1$ and delay size sequence $\{d_{a \geq 0}\}$ with

Fig. 3.14 Worst case state and input trajectories of the CSTR of Eqs. 2.62–2.63 under the two-tier control architecture with the networked LMPC of Eqs. 3.66–3.73

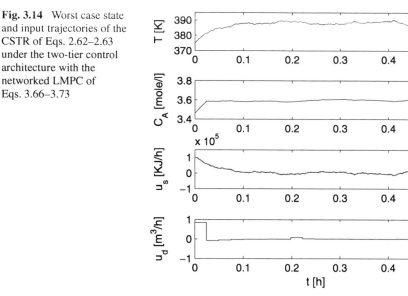

simulation length of 0.5 h are the following:

$$\{t_{a \geq 0}\} = \{0\ 0.198\ 0.395\ 0.500\}\ \text{h}, \tag{3.92}$$

$$\{d_{a \geq 0}\} = \{0\ 0.150\ 0.150\ 0.150\}\ \text{h}. \tag{3.93}$$

The state and input trajectories of the CSTR under the two-tier control architecture with the upper-tier LMPC of Eqs. 3.66–3.73 are shown in Fig. 3.14. From Fig. 3.14, we see that the two-tier control architecture stabilizes the temperature and concentration of the system at the desired equilibrium point in about 0.1 h and 0.05 h, respectively. This implies that the resulting closed-loop system response is faster for this particular simulation. Moreover, the cost associated with the resulting closed-loop trajectories is lower. This result has been validated by extensive simulations.

We also carried out a set of simulations to compare the two-tier control architecture with the lower-tier PI control system from a performance point of view. Table 3.8 shows the total cost computed for 20 different closed-loop simulations under the two-tier control architecture with the LMPC of Eqs. 3.66–3.73 and the PI controller. To carry out this comparison, we have computed the total cost of each simulation based on the performance index defined in Eq. 3.51 from the initial time to the end of the simulation $t_f = 0.5$ h. For each pair of simulations (one for each control scheme), a different initial state inside the stability region, a different uncertainty trajectory and a different random concentration measurement sequence with random delay size sequence are generated. As it can be seen in Table 3.8, the two-tier control architecture has a cost lower than the corresponding total cost under the PI controller in all the closed-loop system simulations.

Table 3.8 Total performance costs along the closed-loop trajectories of the CSTR of Eqs. 2.62–2.63 under the PI controller of Eq. 3.48 and the two-tier control architecture with the upper-tier LMPC of Eqs. 3.66–3.73

sim.	Two-Tier	PI	sim.	Two-Tier	PI
1	107.60	557.06	2	124.98	1090.29
3	188.53	1392.73	4	169.06	403.82
5	143.07	376.15	6	179.22	1330.25
7	202.28	1252.36	8	152.23	749.93
9	141.84	732.20	10	157.99	1049.38

Table 3.9 Steady-state values of manipulated inputs of the reactor–separator process of Eqs. 3.52–3.63

Parameters	Values
Q_{1s}	12.6×10^5 [KJ/hr]
Q_{2s}	16.2×10^5 [KJ/hr]
Q_{3s}	12.6×10^5 [KJ/hr]
F_{20s}	5.04 [m^3/hr]

3.6.4 Application to a Reactor–Separator Process

Consider the reactor–separator process of Eqs. 3.52–3.63 introduced in Sects. 1.2.3 and 3.5.4 with the parameter values given in Table 3.3. We assume that the measurements of temperatures T_1, T_2 and T_3 are available continuously, and the measurements of mass fractions x_{A1}, x_{B1}, x_{A2}, x_{B2}, x_{A3} and x_{B3} are available asynchronously at time instants $\{t_{a \geq 0}\}$ and are subject to time-varying measurement delay. We also assume that there exists a lower bound $T_{\min} = 0.2$ h on the time interval between two consecutive measurements of the mass fractions. The same method used in the previous examples in this chapter is used in the present example to generate the time sequence $\{t_{a \geq 0}\}$. The control objective is to steer the system from the initial state:

$$x(0)^T = [0.890, 0.110, 388.7, 0.886, 0.113, 386.3, 0.748, 0.251, 390.6],$$
$$(3.94)$$

to the steady-state:

$$x_s^T = [0.383, 0.581, 447.8, 0.391, 0.572, 444.6, 0.172, 0.748, 449.6], \quad (3.95)$$

corresponding to the operating condition shown in Table 3.9.

In the present example, we assume that $y_s^T = [y_{s1} \, y_{s2} \, y_{s3}] = [x_3 \, x_6 \, x_9]$ is obtained from the continuous temperature measurements and $y_a^T = [x_1 \, x_2 \, x_4 \, x_5 \, x_7 \, x_8]$ is obtained from the sampled asynchronous, delayed mass fraction measurements. We use the same performance index defined in Eq. 3.51 with Q_c given in Eq. 3.65. We also use the same lower-tier PI controllers as used in Sect. 3.5.4 which are designed based on the continuous temperature measurements (i.e., $y_s(t)$). The same PI controller parameters and Lyapunov function $V(x)$ as in the example of Sect. 3.5.4 are also used.

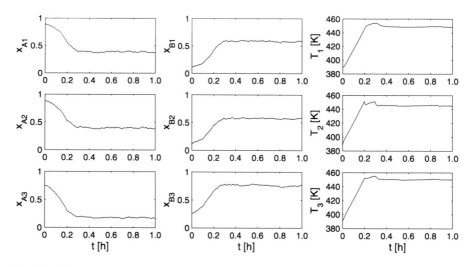

Fig. 3.15 State trajectories of the reactor–separator process of Eqs. 3.52–3.63 under the two-tier control architecture with the networked LMPC of Eqs. 3.66–3.73

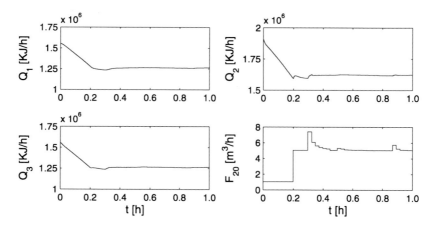

Fig. 3.16 Input trajectories of the reactor–separator process of Eqs. 3.52–3.63 under the two-tier control architecture with the networked LMPC of Eqs. 3.66–3.73

We design the upper-tier LMPC of Eqs. 3.66–3.73 based on the three PI controllers. The feed flow rate to vessel 2, $u_a = F_{20} - F_{20s}$, is the manipulated input for the LMPC, which is bounded by $1 \leq F_{20} \leq 9$ m^3/h. The sampling time of the LMPC is chosen to be $\Delta = 0.025$ h; the prediction horizon is chosen to be $N = 8$. For the simulations carried out in this subsection, we set the prediction horizon $N\Delta$ to be equal to the minimum time interval between two consecutive y_a measurements, T_{min}, and the delay associated with each measurement to be $d_a = D = 0.2$ h for all a which also corresponds to the worst-case effect of measurement delays.

sim.	Two-Tier	PI
1	1.006×10^4	2.148×10^4
2	2.046×10^4	3.123×10^4
3	3.621×10^4	6.310×10^4
4	1.148×10^4	4.440×10^4
5	3.103×10^4	6.052×10^4
6	7.141×10^4	1.631×10^5
7	1.389×10^4	6.961×10^4
8	1.928×10^4	2.770×10^4
9	1.872×10^4	8.538×10^4
10	1.417×10^4	7.260×10^4

Table 3.10 Total performance costs along the closed-loop trajectories of the reactor–separator process of Eqs. 3.52–3.63 under the PI controller and the two-tier control architecture with the upper-tier LMPC of Eqs. 3.66–3.73

The mass fraction measurement sequence $\{t_{a\geq0}\}$ (generated with $W = 1$) and the delay size sequence $\{d_{a\geq0}\}$ with a simulation length 0.75 h are shown below:

$$\{t_{a\geq0}\} = \{0, 0.248, 0.495, 0.868, 1.000\} \text{ h}, \tag{3.96}$$

$$\{d_{a\geq0}\} = \{0, 0.200, 0.200, 0.200, 0.200\} \text{ h}. \tag{3.97}$$

The state and input trajectories of the reactor–separator process of Eqs. 3.52–3.63 under the two-tier control architecture with the upper-tier LMPC of Eqs. 3.66–3.73 are shown in Figs. 3.15 and 3.16. Figure 3.15 shows that the two-tier control architecture drives the temperatures and the mass fractions in the closed-loop system close to the equilibrium point in about 0.25 h. This implies that the resulting closed-loop system response is faster relative to the speed of the closed-loop response under the lower-tier PI controllers. For the same simulation length of $t_f = 1$ h, the performance cost associated with the resulting closed-loop trajectories is 8.658×10^4 which is much smaller than that of the closed-loop system under the lower-tier PI control system (2.105×10^5).

Moreover, we carried out a set of simulations to compare the two-tier control architecture with the lower-tier PI control system with the same parameters from a performance point of view. Table 3.10 shows the total cost computed for 10 different closed-loop simulations under the two-tier control architecture and the lower-tier PI control system. To carry out this comparison, we have computed the total cost of each simulation based on the performance index defined in Eq. 3.51 with different operating conditions. The length of each simulation is $t_f = 0.75$ h. For this set of simulations, W is chosen to be 1. For each pair of simulations (one for each control scheme), a different initial state inside the stability region, a different noise trajectory and a different random mass fraction measurement sequence with random delay size sequence are generated. As can be seen in Table 3.10, the two-tier control architecture has a cost lower than the corresponding total cost under the lower-tier PI control system in all the simulations.

Finally, we studied the effect of input constraints on the performance of the closed-loop system under the two-tier control architecture. Specifically, in this set

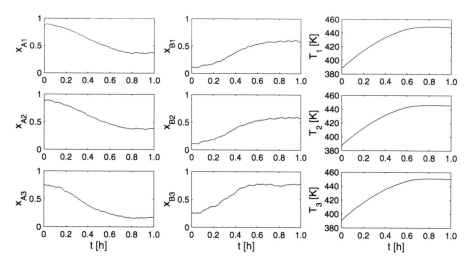

Fig. 3.17 State trajectories of the reactor–separator process of Eqs. 3.52–3.63 subject to input constraints under the lower-tier PI controller

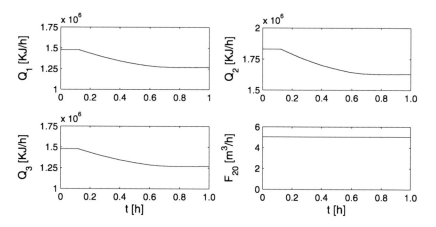

Fig. 3.18 Input trajectories of the reactor–separator process of Eqs. 3.52–3.63 subject to input constraints under the lower-tier PI controller

of simulations, we take into account input constraints in the lower-tier controller manipulated inputs u_s, namely $|Q_1| \leq 1.48 \times 10^5$ KJ/h, $|Q_2| \leq 1.83 \times 10^5$ KJ/h and $|Q_3| \leq 1.48 \times 10^5$ KJ/h. The same simulation settings (initial condition, target state, lower-tier controller design, upper-tier controller design, mass fraction measurement sequence and delay size sequence) as in the previous simulations are used.

The state and input trajectories under the lower-tier PI controllers are shown in Figs. 3.17 and 3.18. From Fig. 3.17, we see that the PI controllers stabilize the system at the target steady-state in about 0.8 h which is a little slower than the corresponding closed-loop response without input constraints (in such a case the closed-loop system is stabilized in about 0.7 h). From Fig. 3.18, we see that the

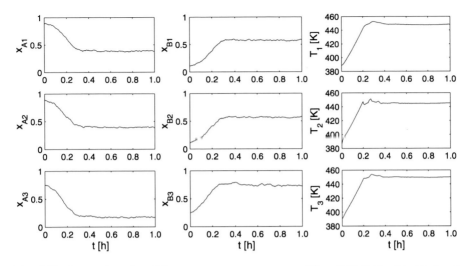

Fig. 3.19 State trajectories of the reactor–separator process of Eqs. 3.52–3.63 subject to input constraints under the two-tier control architecture with the networked LMPC of Eqs. 3.66–3.73

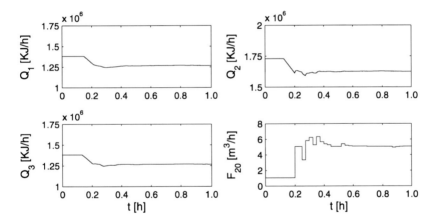

Fig. 3.20 Input trajectories of the reactor–separator process of Eqs. 3.52–3.63 subject to input constraints under the two-tier control architecture with the networked LMPC of Eqs. 3.66–3.73

three heat inputs Q_1, Q_2 and Q_3 operate at their maximum allowable values for about 0.15 h. The corresponding accumulated performance cost is 2.180×10^5.

The state and input trajectories under the two-tier control architecture with the upper-tier LMPC of Eqs. 3.66–3.73 are shown in Figs. 3.19 and 3.20. Figure 3.19 shows that the two-tier control architecture drives the temperatures and the mass fractions of the closed-loop system close to the equilibrium point in about 0.3 h which is a little slower than the closed-loop system response without input constraints (in this case the closed-loop system stabilizes in about 0.25 h). From Fig. 3.20, we see that the heat inputs Q_1, Q_2 and Q_3 also operate at their max-

imum allowable values for about 0.15 h. The corresponding accumulated performance cost is 9.443×10^4 which is much smaller than the cost obtained under the
lower-tier control system (2.180×10^5). From this set of simulations, we see that the
two-tier control architecture maintains the property of improving the performance
of the closed-loop system when input constraints are present. It is also important
to note that advanced anti-windup schemes could be used in conjunction with the
lower-tier PI controller to mitigate the effect of integrator wind-up and improve the
closed-loop system performance; however, the basic conclusion of this part of the
study would not change.

Remark 3.18 In some applications, when input constraints are present, the stability of the closed-loop system under the lower-tier controller may be lost because
of saturation of the control inputs. To avoid loosing stability, the lower-tier controller in the two-tier control architecture can be detuned to primarily take care of
the closed-loop system stability by sacrificing closed-loop performance. Thus, when
input constraints are present, the lower-tier controller can be potentially detuned to
satisfy the input constraints (or saturate for less time) and the upper-tier controller
can be used to recover the loss of closed-loop performance.

3.7 Application to a Wind–Solar Energy Generation System

In this section, we apply the two-tier control architecture to develop a supervisory
predictive control method for the optimal management and operation of a wind–
solar energy generation system. We design a supervisory control system via MPC
which computes the power references for the wind and solar subsystems at each
sampling time while minimizing a suitable cost function. The power references are
sent to two local controllers which drive the wind and solar subsystems to the desired power reference values. We discuss how we can incorporate practical considerations (for example, how to extend the life time of the equipments by reducing
the peak values of inrush or surge currents) into the formulation of the MPC optimization problem by determining an appropriate cost function and constraints. We
will present several simulation case studies that demonstrate the applicability and
effectiveness of the proposed supervisory predictive control architecture.

3.7.1 Wind–Solar System Description

The wind–solar energy generation system considered in this section is based on the
models developed in [104–106]. A schematic of the system is shown in Fig. 3.21.
In this system, there are three subsystems: wind subsystem, solar subsystem and a
lead-acid battery bank which is used to overcome periods of scarce generation.

First, we describe the modeling of the wind subsystem. In the wind energy generation subsystem, there is a windmill, a multipolar permanent-magnet synchronous

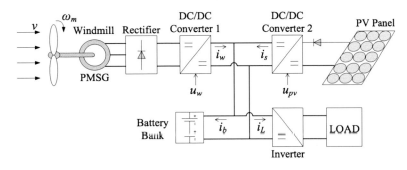

Fig. 3.21 Wind–solar energy generation system

generator (PMSG), a rectifier, and a DC/DC converter to interface the generator
with the DC bus. The converter is used to control indirectly the operating point of
the wind turbine (and consequently its power generation) by commanding the voltage on the PMSG terminals.

The mathematic description of the wind subsystem written in a rotor reference
frame is as follows [105]:

$$\dot{i}_q = -\frac{R_s}{L}i_q - \omega_e i_d + \frac{\omega_e \phi_m}{L} - \frac{\pi v_b i_q u_w}{3\sqrt{3}L\sqrt{i_q^2 + i_d^2}}, \tag{3.98}$$

$$\dot{i}_d = -\frac{R_s}{L}i_d - \omega_e i_q - \frac{\pi v_b i_d u_w}{3\sqrt{3}L\sqrt{i_q^2 + i_d^2}}, \tag{3.99}$$

$$\dot{\omega}_e = \frac{P}{2J}\left(T_t - \frac{3}{2}\frac{P}{2}\phi_m i_q\right), \tag{3.100}$$

where i_q and i_d are the quadrature current and the direct current in the rotor reference frame, respectively; R_s and L are the per phase resistance and inductance of
the stator windings, respectively; ω_e is the electrical angular speed; ϕ_m is the flux
linked by the stator windings; v_b is the voltage on the battery bank terminals; u_w
is the control signal (duty cycle of the DC/DC converter (DC/DC Converter 1 in
Fig. 3.21)), P is the PMSG number of poles, J is the inertial of the rotating parts
and T_t is the wind turbine torque. The wind turbine torque can be written as:

$$T_t = \frac{1}{2}C_t(\lambda)\rho A R v^2, \tag{3.101}$$

where ρ is the air density, A is the turbine-swept area, R is the turbine radius, v is
the wind speed, and $C_t(\lambda)$ is a nonlinear torque coefficient which depends on the tip
speed ratio ($\lambda = \frac{R\omega_m}{v}$ with $\omega_m = \frac{2\omega_e}{P}$ being the angular shaft speed).

Based on Eqs. 3.98–3.100, we can express the power generated by the wind
subsystem and injected into the DC bus as follows:

$$P_w = \frac{\pi v_b}{2\sqrt{3}}\sqrt{i_q^2 + i_d^2}u_w. \tag{3.102}$$

The model of the wind subsystem can be rewritten in the following compact form:

$$\dot{x}_w = f_w(x_w) + g_w(x_w)u_w, \tag{3.103}$$

where $x_w = [i_q \, i_d \, \omega_e]^T$ is the state vector of the wind subsystem and $f_w = [f_{w1} \, f_{w2} \, f_{w3}]^T$, $g_w = [g_{w1} \, g_{w2} \, g_{w3}]^T$ are nonlinear vector functions whose explicit form is omitted for brevity.

Next, we describe the modeling of the solar subsystem. In the solar subsystem, there is a photo-voltaic (PV) panel array and a half-bridge buck DC/DC converter. The solar subsystem is connected to the DC bus via the DC/DC converter. In this subsystem, similar to the wind subsystem, the converter is used to control the operating point of the PV panels.

The mathematic description of the solar subsystem is as follows [106]:

$$\dot{v}_{pv} = \frac{i_{pv}}{C} - \frac{i_s}{C}u_{pv}, \tag{3.104}$$

$$\dot{i}_s = -\frac{v_b}{L_c} + \frac{v_{pv}}{L_c}u_{pv}, \tag{3.105}$$

$$i_{pv} = n_p I_{ph} - n_p I_{rs}\left(e^{\frac{q(v_{pv}+i_{pv}R_s)}{n_s A_c KT}} - 1\right), \tag{3.106}$$

where v_{pv} is the voltage level on the PV panel array terminals, i_s is the current injected on the DC bus, C and L_c are electrical parameters of the buck converter (DC/DC Converter 2 in Fig 3.21), u_{pv} is the control signal (duty cycle), i_{pv} is the current generated by the PV array, n_s is the number of PV cells connected in series, n_p is the number of series strings in parallel, K is the Boltzman constant, A_c is the cell deviation from the ideal p–n junction characteristic, I_{ph} is the photocurrent, and I_{rs} is the reverse saturation current. The power injected by the PV solar module into the DC bus can be computed by:

$$P_s = i_s v_b. \tag{3.107}$$

Note that this power indirectly depends on the control signal u_{pv}.

The model of the solar subsystem can be rewritten in the following compact form:

$$\dot{x}_s = f_s(x_s) + g_s(x_s)u_{pv}, \tag{3.108}$$

$$h_s(x_s) = 0, \tag{3.109}$$

where $x_s = [v_{pv} \, i_s]^T$ is the state vector of the solar subsystem and $f_s = [f_{s1} \, f_{s2}]^T$, $g_s = [g_{s1} \, g_{s2}]^T$ are nonlinear vector functions and $h_s(x_s)$ is a nonlinear scalar function whose explicit form is omitted for brevity.

The DC bus collects the energy generated by both wind and solar subsystems and delivers it to the load and, if necessary, to the battery bank. The voltage of the DC bus is determined by the battery bank which comprises of lead-acid batteries.The load could be an AC or a DC load. In the case under consideration in this section, it is assumed to be an AC load; therefore, a voltage inverter is required. We also assume that the future load of the system for certain length of time is known, that is the total power demand is known.

Because all subsystems are linked to the DC bus, their concurrent effects can be easily analyzed by considering their currents in the common DC side. In this way, assuming an ideal voltage inverter, the load current can be referred to the DC side as an output variable current i_L. Therefore, the current across the battery bank can be written as:

$$i_b = \frac{\pi}{2\sqrt{3}}\sqrt{i_q^2 + i_d^2}u_w + i_s - i_L, \tag{3.110}$$

where i_L is assumed to be a known current.

The lead-acid battery bank may be modeled as a voltage source E_b connected in series with a resistance R_b and a capacitance C_b. Based on this simple model and Eq. 3.110, the DC bus voltage expression can be written as follows:

$$v_b = E_b + v_c + \left(\frac{\pi}{2\sqrt{3}}\sqrt{i_q^2 + i_d^2}u_w + i_s - i_L\right)R_b, \tag{3.111}$$

where v_c is the voltage in capacitor C_b and its dynamics can be described as follows:

$$\dot{v}_c = \frac{1}{C_b}\left(\frac{\pi}{2\sqrt{3}}\sqrt{i_q^2 + i_d^2}u_w + i_s - i_L\right). \tag{3.112}$$

The model of the battery bank can also be rewritten in the following compact form:

$$\dot{v}_c = f_c(x_w, x_s, v_c), \tag{3.113}$$

where $f_c(x_w, x_s, v_c)$ is a nonlinear scalar function.

The dynamics of the generation system can be written in the following compact form:

$$\dot{x} = f(x) + g(x)u, \tag{3.114}$$

$$h(x) = 0, \tag{3.115}$$

where $x = [x_w^T\, x_s^T\, v_c]$, $u = [u_w\, u_{pv}]$, $f(x)$ and $g(x)$ are suitable composition of f_w, f_s, g_w, g_s and f_c, and $h(x) = h_s(x_s)$. The explicit forms of $f(x)$ and $g(x)$ are omitted for brevity.

Note that the maximum power that can be drawn from the wind and solar subsystems is determined by the maximum power that can be generated by the two subsystems. When the two subsystems are not sufficient to complement the generation to satisfy the load requirements, the battery bank can discharge to provide extra power to satisfy the load requirements. However, when the power limit that can be provided by the battery bank is surpassed, the load must be disconnected to recharge the battery bank and avoid damages. In this section, we do not consider the power needed to charge the battery bank explicitly. However, this power can be lumped

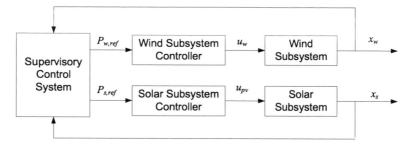

Fig. 3.22 Supervisory control of a wind–solar energy generation system

into the total power demand. In the reminder of this section, we refer to the total power demand as P_T.

3.7.2 Control Problem Formulation and Controller Design

We consider two control objectives of the wind–solar energy generation system. The first and primary control objective is to compute the operating points of the wind subsystem and of the solar subsystem together to generate enough energy to satisfy the load demand. The second control objective is to optimize the operating points to reduce the peak value of surge currents. With respect to the second control objective, specifically, we consider that there are maximum allowable increasing rates of the generated power of the two subsystems and that frequent discharge and charge of the battery bank should be avoided to maximize battery life. Note that the constraints on the maximum increasing rates impose indirect bounds on the peak values of inrush or surge currents to the two subsystems.

The control system is shown in Fig. 3.22 in which the supervisory control system optimizes the power references $P_{w,\text{ref}}$ and $P_{s,\text{ref}}$ (operating points) of the wind and solar subsystems, respectively. The two local controllers (wind subsystem controller and the solar subsystem controller) manipulate u_w and u_{pv} to track the power references, respectively.

Remark 3.19 Note that we consider wind–solar energy generation systems that already operate in normal generating conditions, and do not address the issues related to system startup or shut down. Moreover, we focus on the application of the supervisory control system and do not provide specific conditions (and detailed theoretical derivation) under which the stability of the closed-loop system is guaranteed. We also note that, in the case of an energy generation system containing several solar and wind subsystems, the supervisory control approach can be extended to control the system in a conceptually straightforward manner by letting the supervisory controller determine the power references of all the subsystems.

3.7.2.1 Wind Subsystem Controller Design

For the wind subsystem controller, the objective is to track the power reference computed by the supervisory predictive controller.

In order to proceed, we introduce the maximum power that can be provided by a wind subsystem, $P_{w,\max}$, first. $P_{w,\max}$ depends on a few turbine parameters and on a simple measurement of the angular shaft speed as follows [105]:

$$P_{w,\max} = P_{w,\max}(x) = K_{\text{opt}}\omega_m^3 - \frac{3}{2}(i_q^2 + i_d^2)r_s, \tag{3.116}$$

where $K_{\text{opt}} = \frac{C_t(\lambda_{\text{opt}})\rho A R^3}{2\lambda_{\text{opt}}^2}$ and λ_{opt} is the tip speed ratio at which the coefficient $C_p(\lambda) = C_t(\lambda)\lambda$ reaches its maximum [105], and $C_t(\cdot)$ is the torque coefficient of the wind turbine.

We follow the controller design proposed in [107]. Specifically, the controller is designed as follows:

$$u_w = \begin{cases} u_{w1} & \text{if } P_{w,\text{ref}} < P_{w,\max}, \\ u_{w2} & \text{if } P_{w,\text{ref}} \geq P_{w,\max}, \end{cases} \tag{3.117}$$

where:

$$\begin{aligned} u_{w1} = &-\big[6r_s(i_q f_{w1} + i_d f_{w2}) - 3\phi_{sr}(\omega_e f_{w1} + i_q f_{w3}) + 2\big(\gamma \|s_{w1}(x_w)\| \\ &+ \xi_{\max}\|\partial s_{w1}/\partial x_w\|\big)\text{sign}\big(s_{w1}(x_w)\big)\big] \\ &/\big(6r_s(i_q g_{w1} + i_d g_{w2}) - 3\phi_{sr}\omega_e g_{w1}\big) \end{aligned} \tag{3.118}$$

and

$$\begin{aligned} u_{w2} = &-f_{w1}/g_{w1} + 2K_{\text{opt}}\omega_e f_{w3}/(\phi_{sr}g_{w1}) - i_q f_{w3}/(g_{w1}\omega_e) + 2\big(\gamma \|s_{w2}(x_w)\| \\ &+ \xi_{\max}\|\partial s_{w2}/\partial x_w\|\big)\text{sign}\big(s_{w2}(x_w)\big)/(3\phi_{sr}\omega_e g_{w1}) \end{aligned} \tag{3.119}$$

with $\gamma = 1000$ and $\xi_{\max} = 0.02$ being design constants and

$$\left\|\frac{\partial s_{w1}}{\partial x_w}\right\| = \frac{3}{2}\sqrt{4r_s^2(i_q^2 + i_d^2) + \phi_{sr}^2(\omega_e^2 + i_q^2) - 4r_s\phi_{sr}\omega_e i_q} \tag{3.120}$$

and

$$\left\|\frac{\partial s_{w2}}{\partial x_w}\right\| = \sqrt{\left(\frac{3}{2}\phi_{sr}\omega_e\right)^2 + \left(3K_{\text{opt}}\omega_e^2 - \frac{3}{2}\phi_{sr}i_q\right)^2}. \tag{3.121}$$

In the control design shown in Eq. 3.117, $s_{w1} = P_{w,\text{ref}} - P_w$ and $s_{w2} = P_{w,\max}$ are the sliding surfaces. When the power reference is less than the maximum power that can be provided by the wind subsystem, the control law u_{w1} will operate the subsystem to generate the desired power; when the power reference is greater than the maximum power that can be provided by the wind subsystem, the control law u_{w2} will drive the subsystem to operate at points in which the subsystem provides the maximum power.

3.7.2.2 Solar Subsystem Controller Design

The objective of the solar subsystem controller is to force the subsystem to track the power reference computed by the supervisory controller. The maximum power operating point (MPOP) of the solar subsystem can be computed, in principle, by the following expression [106]:

$$\frac{\partial P_{pv}}{\partial v_{pv}} = \frac{\partial i_{pv}}{\partial v_{pv}} v_{pv} + i_{pv} = 0. \tag{3.122}$$

The maximum solar power provided, $P_{pv,\max}$, is computed numerically through direct evaluation of the following expression [106] in the region where Eq. 3.122 is close to zero:

$$P_{pv,\max} = P_{pv,\max}(x) = -\frac{\partial i_{pv}}{\partial v_{pv}} v_{pv}^2 \cong -\frac{\Delta i_{pv}}{\Delta v_{pv}} v_{pv}^2. \tag{3.123}$$

We follow the controller design proposed in [106] to design the solar subsystem controller. Specifically, this controller is designed as follows:

$$\begin{cases} \text{if } P_{pv,\max} \geq P_{s,\text{ref}}, & u_{pv} = \begin{cases} 1 & \text{if } h_1 \geq 0, \\ 0 & \text{if } h_1 < 0, \end{cases} \\ \text{if } P_{pv,\max} < P_{s,\text{ref}}, & u_{pv} = \begin{cases} 0 & \text{if } h_2 \geq 0, \\ 1 & \text{if } h_2 < 0, \end{cases} \end{cases} \tag{3.124}$$

where $h_1 = P_{s,\text{ref}} - i_s v_b$ and $h_2 = \partial i_{pv}/\partial v_{pv} + i_{pv}/v_{pv}$.

3.7.2.3 Supervisory Controller Design

The objective of the supervisory control system is to determine the power references of the wind and solar subsystems. We will design the supervisory controller via MPC. By using MPC, we can take optimality considerations into account as well as handle different kinds of constraints. As stated before, the primary control objective is to manipulate the operating points of the wind subsystem and of the solar subsystem together to generate enough energy to satisfy the load demand. This control objective will be considered in the design of the cost function for the MPC optimization problem (please see Sect. 3.7.3). The second control objective is to optimize the operating points to reduce the peak value of surge currents. In order to take into account this control objective, we will incorporate hard constraints in the MPC optimization problem to restrict the maximum increasing rates of the generated power of the two subsystems as well as a term in the cost function to avoid frequent discharge and charge of the battery bank.

We consider the case where the future load of the system for certain length of time is known, that is the total power demand, $P_T(t)$, is known. The main implementation element of supervisory predictive control is that the supervisory controller is evaluated at discrete time instants $t_k = t_0 + k\Delta$, $k = 0, 1, \ldots$, with t_0 the initial time and Δ the sampling time, and the optimal future power references, $P_{w,\text{ref}}$ and $P_{s,\text{ref}}$, for a time period (prediction horizon) are obtained and only the first part of the references are sent to the local control systems and implemented on the two units. In order to design this controller, first, a proper number of prediction steps, N, and a sampling time, Δ, are chosen.

The MPC design for the supervisory control system is described as follows:

$$\min_{P_{w,\text{ref}}, P_{s,\text{ref}} \in S(\Delta)} \int_{t_k}^{t_{k+N}} L\big(\tilde{x}(\tau), P_{w,\text{ref}}(\tau), P_{s,\text{ref}}(\tau)\big)\, d\tau, \tag{3.125}$$

$$\text{s.t.} \quad P_{w,\text{ref}}(t) \leq \min_t \{P_{w,\max}(t)\}, \quad t \in [t_{k+j}, t_{k+j+1}), \tag{3.126}$$

$$P_{s,\text{ref}}(t) \leq \min_t \{P_{pv,\max}(t)\}, \quad t \in [t_{k+j}, t_{k+j+1}), \tag{3.127}$$

$$P_{w,\text{ref}}(t_{k+j+1}) - P_{w,\text{ref}}(t_{k+j}) \leq dP_{w,\max}, \tag{3.128}$$

$$P_{s,\text{ref}}(t_{k+j+1}) - P_{s,\text{ref}}(t_{k+j}) \leq dP_{s,\max}, \tag{3.129}$$

$$\dot{\tilde{x}}(t) = f\big(\tilde{x}(t)\big) + g\big(\tilde{x}(t)\big)u(t), \tag{3.130}$$

$$h\big(\tilde{x}(t)\big) = 0, \tag{3.131}$$

$$\tilde{x}(t_k) = x(t_k), \tag{3.132}$$

$$P_{w,\max}(t) = P_{w,\max}\big(\tilde{x}(t)\big), \tag{3.133}$$

$$P_{pv,\max}(t) = P_{pv,\max}\big(\tilde{x}(t)\big), \tag{3.134}$$

where $\tilde{x}$ is the predicted future state trajectory of the wind–solar energy generation system, $L(x, P_{w,\text{ref}}, P_{s,\text{ref}})$ is a positive definite function of the state and the two power references that defines the optimization cost, $dP_{w,\max}$ and $dP_{s,\max}$ are the maximum allowable increasing values of $P_{w,\text{ref}}$ and $P_{s,\text{ref}}$ in two consecutive power references, N is the prediction horizon, $j = 0, \ldots, N - 1$ and $x(t_k)$ is the state measurement obtained at time t_k. We denote the optimal solution to the optimization problem of Eqs. 3.125–3.134 as $P^*_{w,\text{ref}}(t|t_k)$ and $P^*_{s,\text{ref}}(t|t_k)$ which are defined for $t \in [t_k, t_{k+N})$.

The power references of the two subsystems generated by the supervisory controller of Eqs. 3.125–3.134 are defined as follows:

$$P_{w,\text{ref}}(t) = P^*_{w,\text{ref}}(t|t_k), \quad \forall t \in [t_k, t_{k+1}), \tag{3.135}$$

$$P_{s,\text{ref}}(t) = P^*_{s,\text{ref}}(t|t_k), \quad \forall t \in [t_k, t_{k+1}). \tag{3.136}$$

In the optimization problem of Eqs. 3.125–3.134, Eq. 3.125 defines the optimization cost that needs to be minimized, which will be carefully designed in the simulations in Sect. 3.7.3. Because the MPC optimizes the two power references in a discrete time fashion and the references are constants within each sampling interval, the constraints of Eqs. 3.126–3.127 require that the computed power references should be smaller than the minimal of the maximum available within each sampling interval, which means the power references should be achievable for the wind and solar subsystems. Constraints of Eqs. 3.128–3.129 impose constraints on the increasing rate of the two power references. In order to estimate the maximum available power of the two subsystems along the prediction horizon, the model of the system (Eq. 3.130), the current state (Eq. 3.131) and the equations expressing the relation between the maximum available power and the state of each subsystem (Eq. 3.133 and Eq. 3.134) are used. Note that in the MPC optimization problem, in order to estimate the future maximum available power of each subsystem, we assume that the environmental conditions such as wind speed, insolation and temperature remain constant. When the sampling time is small enough and the prediction horizon is short enough, along with high-frequency wind variations caused by gusts and turbulence being reasonably neglected, this assumption makes physical sense [104].

In the remainder of this section, the sampling time and the prediction horizon of the MPC are chosen to be $\Delta = 1$ s and $N = 2$. The maximum increasing values of the two power references are chosen to be $dP_{w,\max} = 1000$ W and $dP_{s,\max} = 500$ W, respectively. Note that the choice of the prediction horizon is based on the fast dynamics of the generation system, the uncertainty associated with long-term future power demand and is also made to achieve a balance between the evaluation time of the optimization problem of the supervisory MPC and the desired closed-loop performance.

3.7.3 Simulation Results

In this subsection, we carry out several sets of simulations to demonstrate the effectiveness and applicability of the designed MPC when the control objectives are taken into account. Note that in all the simulations, standard numerical methods, e.g., Runge–Kutta, are used to carry out the numerical integration of the closed-loop system.

3.7.3.1 Constraints on the Maximum Increasing Rates of $P_{w,\text{ref}}$ and $P_{s,\text{ref}}$

In this set of simulations, the control objective is to operate the wind–solar energy generation system to satisfy the total power demand, P_T, subject to constraints on the rate of change of $P_{w,\text{ref}}$ and $P_{s,\text{ref}}$. Because the constraints on the maximum increasing rates of $P_{w,\text{ref}}$ and $P_{s,\text{ref}}$ are considered as hard constraints in the formulation of the MPC (i.e., constraints of Eqs. 3.128–3.129), in the cost function, we only penalize the total power demand. The cost function designed for these control objectives is shown as follows:

$$L(x, P_{w,\text{ref}}, P_{s,\text{ref}}) = \alpha(P_T - P_{w,\text{ref}} - P_{s,\text{ref}})^2 + \beta P_{s,\text{ref}}^2, \qquad (3.137)$$

where $\alpha = 1$ and $\beta = 0.01$ are constant weighting factors. The first term, $\alpha(P_T - P_{w,\text{ref}} - P_{s,\text{ref}})^2$, in the cost function penalizes the difference between the power generated by the wind–solar system and the total power demand, which drives the wind and solar subsystems to satisfy the total demand to the maximum extent. Because there are infinite combinations of $P_{w,\text{ref}}$ and $P_{s,\text{ref}}$ that can minimize the first term, in order to get a unique solution to the optimization problem, we also put a small penalty on $P_{s,\text{ref}}$. This implies that the wind subsystem is operated as the primary generation system and the solar subsystem is only activated when the wind subsystem alone can not satisfy the power demand. In the simulation, we assume that the environmental conditions remain constant with wind speed $v = 12$ m/s, insolation $\lambda_l = 90$ mW/cm^2 and PV panel temperature $T = 65°C$.

Figure 3.23 shows the results of the simulations. From Fig. 3.23, we see that at $t = 4$ s there is a demand power increase from 2100 W to 4000 W (Fig. 3.23(a)), and that because of the constraints on the maximum increasing rates of $P_{w,\text{ref}}$ and $P_{s,\text{ref}}$, the wind–solar system cannot supply sufficient power (Fig. 3.23(b)–(c)) and the shortage of power is made up by the battery bank (Fig. 3.23(a)).

Note that we assume that the future power demand for a short time period is known to the MPC. Because of this, at $t = 8$ s, when the MPC supervisory controller receives information about a power demand increase at $t = 9$ s, and having information of the limits on the power generation of the two subsystems, it coordinates the power generations of the wind and solar subsystems to best satisfy the power demand by reducing the power generation of the wind subsystem and activating the solar subsystem in advance at $t = 8$ s. This coordination renders the two subsystems able to approach as much as possible to the total power demand requirement at $t = 9$ s (even though they cannot fully meet this requirement due to operation constraints of the wind and solar subsystems) by boosting their power production at the maximum possible rate, i.e., about 1,500 W boost in power production from $t = 8$ s to $t = 9$ s. On the other hand, if there is no information of the future power demand increase that is fed to the MPC, the wind–solar system would not increase its production as fast to approach the total power demand requirement because the solar subsystem would stay dormant up to $t = 9$ s (the power demand requirement at $t = 8$ s can be fully satisfied by the wind subsystem only) and the presence of a hard constraint on the rate of change of power generated by the solar subsystem would not allow to boost its production enough to meet the total power demand requirement at $t = 9$ s (in this case, the total power demand requirement cannot be achieved by operation of the wind subsystem only); as a result the boost in total power production in this case would be only 1,200 W.

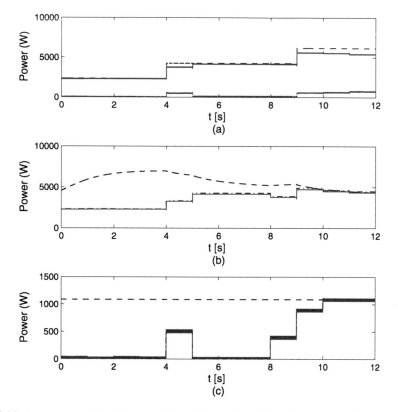

Fig. 3.23 Power trajectories with constraints on the maximum increasing rates of $P_{w,\text{ref}}$ and $P_{s,\text{ref}}$. (a) Generated power $P_w + P_s$ (*solid line*), total power demand P_T (*dashed line*) and power provided by battery bank P_b (*dotted line*); (b) power generated by wind subsystem P_w (*solid line*), wind power reference $P_{w,\text{ref}}$ (*dash-dotted line*) and maximum wind generation $P_{w,\max}$ (*dashed line*); (c) power generated by solar subsystem P_s (*solid line*), solar power reference $P_{s,\text{ref}}$ (*dash–dotted line*) and maximum solar generation $P_{s,\max}$ (*dashed line*)

3.7.3.2 Suppression of Battery Power Fluctuation

In this set of simulations, we modify the cost function of Eq. 3.137 to take into account the fluctuation of the battery power in order to avoid frequent battery charge and discharge. The cost function is modified as follows:

$$L(x, P_{w,\text{ref}}, P_{s,\text{ref}}) = \alpha(P_T - P_{w,\text{ref}} - P_{s,\text{ref}})^2 + \beta P_{s,\text{ref}}^2 + \zeta \Delta P_b^2, \qquad (3.138)$$

where ΔP_b is the change of the power provided by the battery bank between two consecutive steps and $\zeta = 0.4$ is a weighting factor. Note that this newly added term requires that we store the trajectory of P_b. In this set of simulations, the environmental conditions are set with wind speed $v = 11$ m/s, insolation $\lambda_l = 90$ mW/cm^2 and PV panel temperature $T = 65°$C.

Figure 3.24 shows the simulation results. From Fig. 3.24, we see that there is a power demand decrease at $t = 3$ s, and though the wind and solar subsystems are

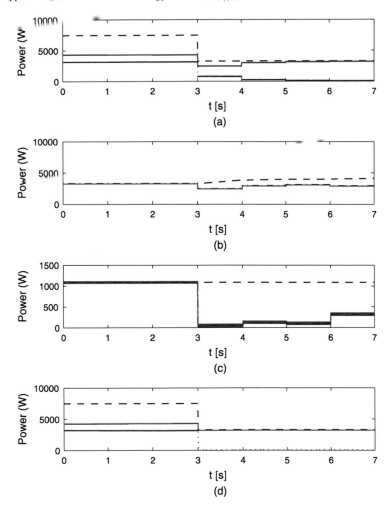

Fig. 3.24 Power trajectories taking into account suppression of battery power fluctuation. (**a**) Generated power $P_w + P_s$ (*solid line*), total power demand P_T (*dashed line*) and power provided by battery bank P_b (*dotted line*); (**b**) power generated by wind subsystem P_w (*solid line*), wind power reference $P_{w,ref}$ (*dash-dotted line*) and maximum wind generation $P_{w,max}$; (**c**) power generated by solar subsystem P_s (*solid line*), solar power reference $P_{s,ref}$ (*dash-dotted line*) and maximum solar generation $P_{s,max}$; (**d**) generated power $P_w + P_s$ (*solid line*), total power demand P_T (*dashed line*) and power provided by battery bank P_b (*dotted line*)

able to provide enough power to satisfy the demand, the supervisory controller will not reduce the power generated by the battery to 0 immediately at $t = 3$ s; instead, the supervisory controller operates the system to make the power provided by the battery bank decrease slower and reach its recharge state at $t = 5$ s (Fig. 3.24(a)). Fig. 3.24(d) shows the power trajectory of the battery bank if no penalty on the change of the power provided by the battery bank is applied.

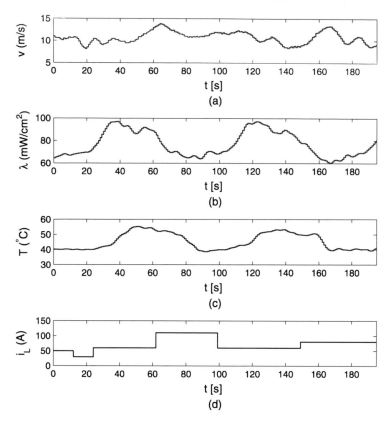

Fig. 3.25 Environmental conditions and load current. (**a**) Wind speed v; (**b**) insolation λ_l; (**c**) PV panel temperature T; (**d**) load current i_L

3.7.3.3 Varying Environmental Conditions

In this part, we carry out simulations under varying environmental conditions. Time evolution of wind speed, PV panel temperature and insolation are shown in Fig. 3.25(a)–(c). Fig. 3.25(d) shows the trajectory of total power demand.

It can be seen from Fig. 3.26(a) that the wind/solar/battery powers coordinate their behavior to meet the load demand. Time evolution of output power and maximum available power from the wind subsystem and solar subsystem are plotted in Fig. 3.26(b)–(c). When sufficient energy supply can be extracted from the two subsystems such as during 0~60 s, 100~140 s and 160~173 s, the battery is being recharged. In other periods, load demand is relatively high and the weather condition, which determines the maximum available generation capacity of the two subsystems, cannot permit sufficient energy supply. Thus, the supervisory controller drives wind/solar parts to their instant maximum capacity and calls the battery bank for shortage compensation.

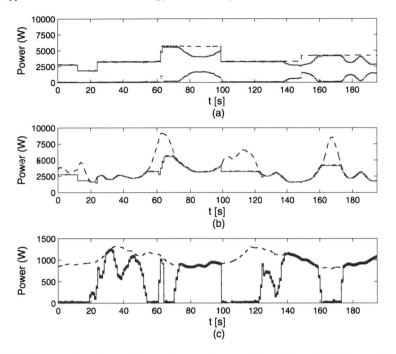

Fig. 3.26 Power trajectories under varying environmental conditions. (**a**) Generated power $P_w + P_s$ (*solid line*), total power demand P_T (*dashed line*) and power provided by battery bank P_b (*dotted line*); (**b**) power generated by wind subsystem P_w (*solid line*), wind power reference $P_{w,\text{ref}}$ (*dash-dotted line*) and maximum wind generation $P_{w,\text{max}}$ (*dashed line*); and (**c**) power generated by solar subsystem P_s (*solid line*), solar power reference $P_{s,\text{ref}}$ (*dash-dotted line*) and maximum solar generation $P_{s,\text{max}}$ (*dashed line*)

3.7.3.4 Consideration of High-Frequency Disturbance of Weather Conditions

In the preceding scenaria, we assumed that the variation of weather-related parameters, like wind speed and insolation, within each sampling time interval is negligible. While this assumption is reasonable in most cases, additional attention for robust system operation should be given under even harsher conditions where high frequency disturbances that influence the values of wind speed and insolation are present. This scenario is possible when the wind turbine encounters turbulent flow [82], or when insolation is affected by abrupt changes in atmospheric turbidity [33].

To study this case from a control point of view and evaluate the robustness of the control system in this case, we introduce disturbances in two parameters; specifically, 10% variation in the wind speed and 5% variation in the insolation. The profiles of the wind speed and insolation are shown in Figs. 3.27(a) and (b). We have used the system model to establish that the control system operating on the wind subsystem can tolerate the wind disturbance and no additional mea-

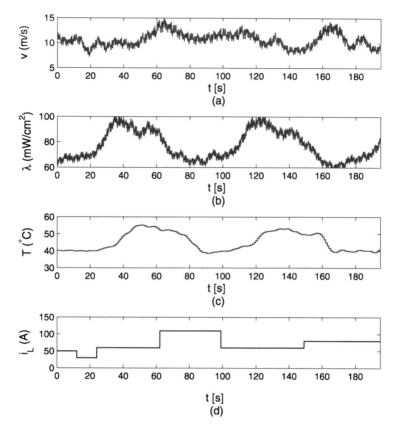

Fig. 3.27 Environmental conditions and load current. (**a**) Wind speed with high frequency disturbance v; (**b**) insolation with high frequency disturbance λ_l; (**c**) PV panel temperature T; (**d**) load
current i_L

sures are needed to be taken to secure its reliability. However, for the solar subsystem, which is characterized by faster dynamics, in order to maintain its closed-
loop stability we need to use a more conservative estimate of the insolation (i.e.,
95% of the value of the measured insolation) in the evaluation of the power reference. This conservative estimate of insolation ensures that the predicted maximum power delivered by the solar subsystem does not exceed what the weather
permits.

The closed-loop profiles of power generation are displayed in Fig. 3.28(a)–(c).
Again, the entire energy generation system operates reliably, thereby yielding positive results for the robustness of the control system with respect to abrupt variations
in wind speed and insolation. Both maximum power generation capabilities of the
two subsystems are perturbed as a result of the weather disturbance, but both the
wind subsystem and the solar subsystem operate in a robust fashion and the total
power demand is met.

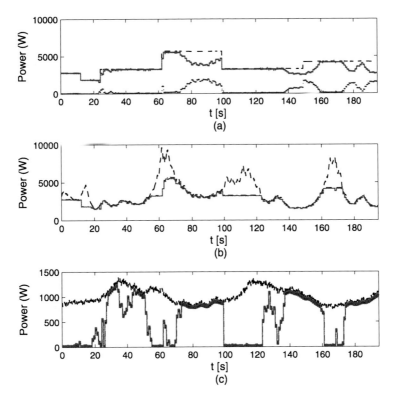

Fig. 3.28 Power trajectories under varying environmental conditions with high frequency disturbance. (**a**) Generated power $P_w + P_s$ (*solid line*), total power demand P_T (*dashed line*) and power provided by battery bank P_b (*dotted line*); (**b**) power generated by wind subsystem P_w (*solid line*), wind power reference $P_{w,ref}$ (*dash-dotted line*) and maximum wind generation $P_{w,max}$ (*dashed line*); and (**c**) power generated by solar subsystem P_s (*solid line*), solar power reference $P_{s,ref}$ (*dash-dotted line*) and maximum solar generation $P_{s,max}$ (*dashed line*)

3.8 Conclusions

In this chapter, we presented a two-tier networked control architecture for process control problems that involve nonlinear processes and heterogeneous measurements consisting of continuous measurements and asynchronous measurements (with or without delays). The presented architecture consists of: (a) a lower-tier control system, which relies on point-to-point communication and continuous measurements, to stabilize the closed-loop system, and (b) an upper-tier networked control system, designed using LMPC theory, that profits from both the continuous and the asynchronous, delayed measurements as well as from additional networked control actuators to improve the closed-loop system performance. The applicability and effectiveness of the methods were demonstrated using two chemical process examples.

In addition, the two-tier control architecture was also applied to the supervisory control of a standalone wind–solar energy generation system. Specifically, we focused on the development of a supervisory predictive control method for the optimal management and operation of wind–solar energy generation systems. We designed a supervisory control system designed via MPC which computes the power references for the wind and solar subsystems at each sampling time while minimizing a suitable cost function. The power references are sent to two local controllers which drive the two subsystems to the power references. We discussed how to incorporate practical considerations, for example, how to reduce the peak values of inrush or surge currents, into the formulation of the MPC optimization problem. Simulation results demonstrated the effectiveness and applicability of the presented approach.

Chapter 4
Distributed Model Predictive Control: Two-Controller Cooperation

4.1 Introduction

In Chap. 3, we presented a two-tier networked control architecture for nonlinear processes, shown in Fig. 3.2. In this architecture, the preexisting local control system (LCS) uses continuous sensing and actuation and an explicit control law (for example, the local controller is a classical controller, like a proportional-integral-derivative controller, or a nonlinear controller designed via geometric or Lyapunov-based control methods for which an explicit formula for the calculation of the control action is available). On the other hand, the networked control system (NCS) uses networked (wired or wireless) sensors and actuators and has access to heterogeneous, asynchronous measurements that are not available to the LCS. The NCS is designed via LMPC. An important feature of the two-tier networked control architecture of Fig. 3.2 is that there is no communication between the LCS and NCS since the networked LMPC can estimate the control actions of the local controller using the explicit formula of this controller, and thus, it can take into account the actions of the local controller in the computation of its optimal input trajectories. In this sense, the two-tier networked control architecture of Fig. 3.2 can be thought of as a decentralized control system. This lack of communication is an appealing feature because the addition of the NCS does not lead to any modification of the preexisting LCS and improves the overall performance and robustness of the combined NCS/LCS architecture (i.e., the achievable closed-loop performance is invariant to disruptions in the communication between the NCS and LCS).

Despite this progress, there are important controller design problems that remain unresolved in the broad context of networked control systems. For example, when the LCS is a model predictive control system for which there is no explicit controller formula to calculate its future control actions, it is necessary to redesign both the NCS and the LCS and establish some level of (preferably small) communication between them so that they can coordinate their actions. To this end, we will adopt in this chapter a distributed MPC (DMPC) approach to the design of the NCS and LCS, as shown in Fig. 4.1. It is important to remark at this point that an alternative approach to address the integrated design of the NCS and LCS would be to design a

P.D. Christofides et al., *Networked and Distributed Predictive Control*,
Advances in Industrial Control,
DOI 10.1007/978-0-85729-582-8_4, © Springer-Verlag London Limited 2011

Fig. 4.1 Distributed MPC
control architecture for
networked control system
design

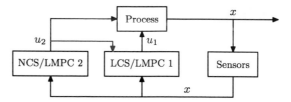

fully centralized MPC to decide the manipulated inputs of all the control actuators (i.e., both u_1 and u_2 in Fig. 4.1). However, the computational complexity of MPC grows significantly with the increase of optimization (decision) variables, which may prohibit certain online centralized MPC applications with a large number of decision variables.

Specifically, in this chapter, we present a DMPC design where both the preexisting local control system and the networked control system are designed via LMPC. The DMPC design that will be presented—see Fig. 4.1—uses a hierarchical control architecture in the sense that the LCS is able to stabilize the closed-loop system and the NCS takes advantage of additional control inputs and coordinates with the LCS to improve the closed-loop performance. This hierarchical DMPC design is different from previous DMPC designs which decompose a centralized control problem spatially (see also Chap. 6 of this book for results in this direction). In particular, the proposed design provides the potential of maintaining stability and performance in the face of new/failing actuators, (for example, the failure of the actuator of the NCS (zero input) in Fig. 4.1 does not affect the closed-loop stability). Working with general nonlinear models of chemical processes and assuming that there exists a nonlinear controller that stabilizes the nominal closed-loop system using only the preexisting control loops (LCS), two separate Lyapunov-based model predictive controllers will be designed that coordinate their actions in an efficient fashion. Specifically, the DMPC design preserves the stability properties of the nonlinear controller, improves the closed-loop performance and allows handling input constraints. In addition, the distributed control design requires reduced communication between the two distributed controllers since it requires that these controllers communicate only once at each sampling time and is computationally more efficient compared to the corresponding centralized MPC design.

In addition, we will extend the results to include nonlinear systems subject to asynchronous and delayed measurements. In the case of asynchronous feedback, under the assumption that there exists an upper bound on the interval between two successive state measurements, distributed controllers that utilize one-directional communication and coordinate their actions to ensure that the state of the closed-loop system is ultimately bounded in a region that contains the origin will be designed. In the case of asynchronous measurements that also involve time-delays, under the assumption that there exists an upper bound on the maximum measurement delay, a DMPC system that utilizes bidirectional communication between the distributed controllers and takes the measurement delays explicitly into account to enforce practical stability in the closed-loop system will be designed. These DMPC designs also possess explicitly characterized sets of initial conditions starting from where they are guaranteed to be feasible and stabilizing. The theoretical results will be demonstrated through a chemical process example. The results of this chapter were first presented in [10, 52, 56].

4.2 System Description

In this chapter, we consider nonlinear systems described by the following state-space model:

$$\dot{x}(t) = f\big(x(t), u_1(t), u_2(t), w(t)\big), \tag{4.1}$$

where $x(t) \in R^n$ denotes the vector of state variables, $u_1(t) \in R^{m_1}$ and $u_2(t) \in R^{m_2}$ are two different sets of control inputs and $w(t) \in R^w$ denotes the vector of disturbance variables. The two sets of control inputs are restricted to be in two nonempty convex sets $U_1 \subseteq R^{m_1}$ and $U_2 \subseteq R^{m_2}$ and the disturbance vector is bounded, i.e., $w(t) \in W$ where:

$$W := \big\{ w \in R^w : \|w\| \le \theta, \theta > 0 \big\} \tag{4.2}$$

with θ being a known positive real number.

We assume that f is a locally Lipschitz vector function and $f(0, 0, 0, 0) = 0$. This means that the origin is an equilibrium point for the nominal system of Eq. 4.1 with $u_1 = 0$ and $u_2 = 0$.

Remark 4.1 In general, distributed control systems are formulated based on the assumption that the controlled systems are decoupled or partially decoupled. However, we consider a fully coupled process model with two sets of possible manipulated inputs; this is a very common occurrence in chemical process control as we will illustrate in the example of Sect. 4.4.3. It is important to note that even though we have motivated the control problem of Eq. 4.1 by the augmentation of LCS with NCS, the same control formulation could be used when a new control system which may use a local control network is added to a process that already operates under an MPC; see example in Sect. 4.4.3.

4.3 Lyapunov-Based Control

We assume that there exists a nonlinear state feedback control law $u_1(t) = h(x(t))$ which satisfies the input constraint on u_1 for all x inside a given stability region and renders the origin of the nominal closed-loop system asymptotically stable with $u_2(t) = 0$. Using converse Lyapunov theorems, this assumption implies that there exist functions $\alpha_i(\cdot)$, $i = 1, 2, 3, 4$ of class $\mathcal{K}$ and a continuous differentiable Lyapunov function $V(x)$ for the nominal closed-loop system that satisfy the following inequalities:

$$\alpha_1\big(\|x\|\big) \le V(x) \le \alpha_2\big(\|x\|\big), \tag{4.3}$$

$$\frac{\partial V(x)}{\partial x} f\left(x, h(x), 0, 0\right) \leq -\alpha_3\left(\|x\|\right), \tag{4.4}$$

$$\left\|\frac{\partial V(x)}{\partial x}\right\| \leq \alpha_4\left(\|x\|\right), \tag{4.5}$$

$$h(x) \in U_1 \tag{4.6}$$

for all $x \in O \subseteq R^n$ where O is an open neighborhood of the origin. We denote the region $\Omega_\rho \subseteq O$ as the stability region of the closed-loop system under the control $u_1 = h(x)$ and $u_2 = 0$.

By continuity, the local Lipschitz property of the vector field $f(x, u_1, u_2, w)$ and the fact that the manipulated inputs u_1 and u_2 are bounded in convex sets, there exists a positive constant M such that:

$$\left\|f(x, u_1, u_2, w)\right\| \leq M \tag{4.7}$$

for all $x \in \Omega_\rho$, $u_1 \in U_1$, $u_2 \in U_2$ and $w \in W$. In addition, by the continuous differentiable property of the Lyapunov function V, there exist positive constants L_x, L_w and L'_x such that:

$$\left\|f(x, u_1, u_2, w) - f\left(x', u_1, u_2, 0\right)\right\| \leq L_w\|w\| + L_x\|x - x'\|, \tag{4.8}$$

$$\left\|\frac{\partial V(x)}{\partial x} f(x, u_1, u_2, w) - \frac{\partial V(x')}{\partial x} f\left(x', u_1, u_2, 0\right)\right\|$$

$$\leq L'_w\|w\| + L'_x\|x - x'\| \tag{4.9}$$

for all $x, x' \in \Omega_\rho$, $u_1 \in U_1$, $u_2 \in U_2$ and $w \in W$. These constants will be used to characterize the stability properties of the system of Eq. 4.1 under the DMPC designs.

4.4 DMPC with Synchronous Measurements

In this section, we present a DMPC design for the system of Eq. 4.1 with synchronous measurements. In Sects. 4.5 and 4.6, we will extend the results presented in this section to include systems subject to asynchronous measurements without and with time-varying delays, respectively.

Specifically, in the current section, we assume that measurements of the system state x are available at synchronous sampling times $\{t_{k\geq 0}\}$ with $t_k = t_0 + k\Delta$, $k = 0, 1, \ldots$ where t_0 is the initial time and Δ is the sampling time.

4.4.1 DMPC Formulation

In this section, we design two separate LMPCs to compute u_1 and u_2 and refer to the LMPC computing the trajectories of u_1 and u_2 as LMPC 1 and LMPC 2, respectively. Figure 4.1 shows a schematic of the distributed control method discussed in this section. The implementation strategy of the distributed control architecture is as follows:

1. At t_k, both LMPC 1 and LMPC 2 receive the state measurement $x(t_k)$ from the sensors.
2. LMPC 2 evaluates the optimal input trajectory of u_2 based on the $x(t_k)$ and sends the first step input value of u_2 (i.e., $u_2 \in [t_k, t_{k+1})$) to its corresponding actuators and the entire optimal input trajectory of u_2 (i.e., $u_2 \in [t_k, t_{k+N})$ with N the prediction horizon of the LMPC3) to LMPC 1.
3. Once LMPC 1 receives the entire optimal input trajectory of u_2 from LMPC 2, it evaluates the future input trajectory of u_1 based on $x(t_k)$ and the entire optimal input trajectory of u_2.
4. LMPC 1 sends the first step input value of u_1 (i.e., $u_1 \in [t_k, t_{k+1})$) to its corresponding actuators.
5. When a new measurement is received ($k \leftarrow k + 1$), go to Step 1.

First, we define the optimization problem of LMPC 2. This optimization problem depends on the latest state measurement $x(t_k)$, however, LMPC 2 does not have any information about the value that u_1 will take. In order to make a decision, LMPC 2 must assume a trajectory for u_1 along the prediction horizon. To this end, the nonlinear control law $u_1 = h(x)$ is used. In order to inherit the stability properties of this control law, u_2 must satisfy a Lyapunov-based constraint that guarantees a given minimum decrease rate of the Lyapunov function V. The design of LMPC 2 is based on the following optimization problem:

$$\min_{u_2 \in S(\Delta)} \int_{t_k}^{t_{k+N}} \left[\left\| \tilde{x}(\tau) \right\|_{Q_c} + \left\| u_1(\tau) \right\|_{R_{c1}} + \left\| u_2(\tau) \right\|_{R_{c2}} \right] d\tau, \tag{4.10}$$

$$\text{s.t.} \quad \dot{\tilde{x}}(t) = f\big(\tilde{x}(t), u_1(t), u_2(t), 0\big), \tag{4.11}$$

$$u_1(t) = h\big(\tilde{x}(t_{k+j})\big), \quad \forall t \in [t_{k+j}, t_{k+j+1}), \; j = 0, \ldots, N-1, \tag{4.12}$$

$$u_2(t) \in U_2, \tag{4.13}$$

$$\tilde{x}(t_k) = x(t_k), \tag{4.14}$$

$$\frac{\partial V(x(t_k))}{\partial x} f\big(x(t_k), h\big(x(t_k)\big), u_2(t_k), 0\big)$$
$$\leq \frac{\partial V(x(t_k))}{\partial x} f\big(x(t_k), h\big(x(t_k)\big), 0, 0\big), \tag{4.15}$$

where Q_c, R_{c1} and R_{c2} are positive definite weighting matrices, $\tilde{x}$ is the predicted trajectory of the nominal system with u_2 being the input trajectory computed by this LMPC and u_1 being the nonlinear control law $h(x)$ applied in a sample-and-hold fashion. The optimal solution to this optimization problem is denoted by $u_2^*(t|t_k)$ which is defined for $t \in [t_k, t_{k+N})$. This information is sent to LMPC 1.

The constraint of Eq. 4.13 defines the constraint on the control input u_2 and the Lyapunov-based constraint of Eq. 4.15 guarantees that the value of the time derivative of the Lyapunov function at the initial evaluation time, if $u_1 = h(x(t_k))$

and $u_2 = u_2^*(t_k|t_k)$ are applied, is lower than or equal to the value obtained when $u_1 = h(x)$ and $u_2 = 0$ are applied.

The optimization problem of LMPC 1 depends on the latest state measurement $x(t_k)$ and the decision taken by LMPC 2 (i.e., $u_2^*(t|t_k)$). This allows LMPC 1 to compute an input u_1 such that the closed-loop performance is optimized, while guaranteeing that the stability properties of the nonlinear control law $h(x)$ are preserved. Specifically, LMPC 1 is based on the following optimization problem:

$$\min_{u_1 \in S(\Delta)} \int_{t_k}^{t_{k+N}} \left[\left\| \tilde{x}(\tau) \right\|_{Q_c} + \left\| u_1(\tau) \right\|_{R_{c1}} + \left\| u_2(\tau) \right\|_{R_{c2}} \right] d\tau, \qquad (4.16)$$

$$\text{s.t.} \quad \dot{\tilde{x}}(t) = f\left(\tilde{x}(t), u_1(t), u_2(t), 0 \right), \qquad (4.17)$$

$$u_1(t) \in U_1, \qquad (4.18)$$

$$u_2(t) = u_2^*(t|t_k), \qquad (4.19)$$

$$\tilde{x}(t_k) = x(t_k), \qquad (4.20)$$

$$\frac{\partial V(x(t_k))}{\partial x} f\left(x(t_k), u_1(t_k), u_2^*(t_k|t_k), 0 \right),$$

$$\leq \frac{\partial V(x(t_k))}{\partial x} f\left(x(t_k), h\left(x(t_k) \right), u_2^*(t_k|t_k), 0 \right), \qquad (4.21)$$

where $\tilde{x}$ is the predicted trajectory of the nominal system with u_2 being the optimal input trajectory $u_2^*(t|t_k)$ computed by LMPC 2 and u_1 being the input trajectory computed by LMPC 1. The optimal solution to this optimization problem is denoted by $u_1^*(t|t_k)$ which is defined for $t \in [t_k, t_{k+N})$.

The constraint of Eq. 4.18 defines the constraint on the control input u_1 and the Lyapunov-based constraint of Eq. 4.21 guarantees that the value of the time derivative of the Lyapunov function at the initial evaluation time, if $u_1 = u_1^*(t_k|t_k)$ and $u_2 = u_2^*(t_k|t_k)$ are applied, is lower than or equal to the value obtained when $u_1 = h(x(t_k))$ and $u_2 = u_2^*(t_k|t_k)$ are applied.

Once both optimization problems are solved, the inputs of the DMPC design based on the above LMPC 1 and LMPC 2 are defined as follows:

$$u_1(t) = u_1^*(t|t_k), \quad \forall t \in [t_k, t_{k+1}), \qquad (4.22)$$

$$u_2(t) = u_2^*(t|t_k), \quad \forall t \in [t_k, t_{k+1}). \qquad (4.23)$$

Remark 4.2 At Step 2 of the presented implementation strategy, the whole optimal input trajectory of LMPC 2 is sent to LMPC 1. From the stability point of view, it is unnecessary to send the whole optimal input trajectory. Only the first step of the optimal input trajectory of LMPC 2 is needed to send to LMPC 1 in order to guarantee the stability of the closed-loop system under the DMPC (please see Sect. 4.4.2 for the proof of the closed-loop stability). Thus, the communication between the two LMPCs can be minimized by only sending the first step of an optimal input trajectory without loss of the closed-loop stability. However, the transmission of the whole optimal trajectory at a sampling time can, to some extend, improve the closed-loop performance because LMPC 1 has more information on the possible future input trajectory of LMPC 2.

Remark 4.3 The key idea of the DMPC formulation is to impose a hierarchy on the order in which the controllers are evaluated in order to guarantee that the resulting control actions stabilize the system. In this section, we assume flawless communications and synchronous state measurements at each sampling time without delay. If data losses and delays are taken into account, the control method has to be modified because at each time step coordination between both LMPCs is not guaranteed; these issues are addressed in Sects. 4.5 and 4.6.

Remark 4.4 Since the computational burden of nonlinear MPC methods is usually high, the DMPC design only requires LMPC 2 and LMPC 1 to "talk" once every sampling time (that is, LMPC 2 sends its optimal input trajectory to LMPC 1) to minimize the communication between the two LMPCs. This strategy is more robust when communication between the distributed MPCs can be subject to disruptions. Note also that the computational complexities of the LMPC optimization problems of the DMPC design can be further reduced by appropriately reducing the dimension of the system model used in the formulation of optimization problems; the reader may refer to [38, 43] for discussion on model reduction via two-time-scale techniques.

Remark 4.5 The constraints of Eqs. 4.15, 4.19 and 4.21 are a key element of the DMPC design. In general, guaranteeing closed-loop stability of a distributed control system is a difficult task because of the interactions between the distributed controllers and can only be done under certain assumptions (see, for example, [8, 92]). The constraint of Eq. 4.19 guarantees that LMPC 1 takes into account the effect of LMPC 2 to the applied inputs (recall that LMPC 2 is designed without taking LMPC 1 into account). The constraints of Eqs. 4.15 and 4.21 together with the hierarchical control strategy (i.e., LMPC 2 is solved first and LMPC 1 is solved second) guarantee that the value of the Lyapunov function of the closed-loop system is a decreasing sequence of time with a lower bound.

Remark 4.6 Note that the stability properties of the closed-loop system are inherited from the nonlinear control law $u_1 = h(x)$. Once the Lyapunov-based constraints of Eqs. 4.15 and 4.21 are satisfied, closed-loop stability is guaranteed. The main purpose of LMPC 1 and LMPC 2 is to optimize the inputs u_1 and u_2 to improve performance. Thus, during the evaluation of the optimal solutions of LMPC 1 and LMPC 2 within a sampling period, we can terminate the optimization (i.e., limit the number of iterations in the process of searching for the optimal solutions) to obtain sub-optimal input trajectories without losing the stability properties. An extreme application of this idea is when the optimization process is terminated at the beginning of every optimization process which gives the inputs: $u_1(t) = h(x(t_k))$ and $u_2(t) = 0$ for $t \in [t_k, t_{k+1})$, which guarantees stability of the closed-loop system but not optimal performance.

Remark 4.7 In the DMPC design of Eqs. 4.10–4.21, LMPC 2 and LMPC 1 are evaluated in sequence, which implies that the minimal sampling time of the system should be greater than or equal to the sum of the evaluation times of LMPC 2 and LMPC 1. In order to solve both optimization problems in parallel, LMPC 1 can use old input trajectories of LMPC 2, that is, at t_k, LMPC 1 uses $u_2^*(t|t_{k-1})$ to define its optimization problem. This strategy, however, may introduce extra errors in the optimization problem and may not guarantee closed-loop stability.

Remark 4.8 The Lyapunov-based constraints of Eqs. 4.10 and 4.21 guarantee that the choice of u_2 cannot render LMPC 1 infeasible. In addition, the two constraints guarantee that the DMPC design inherits the stability region of the nonlinear control law $h(x)$.

4.4.2 Stability Properties

In this subsection, we present the stability properties of the DMPC of Eqs. 4.10–4.21. The DMPC of Eqs. 4.10–4.21 computes the inputs u_1 and u_2 applied to the system of Eq. 4.1 in a way such that the value of the Lyapunov function at time instant t_k (i.e., $V(x(t_k))$) is a decreasing sequence of values with a lower bound. This is achieved due to the Lyapunov-based constraints of Eqs. 4.15 and 4.21. This property is presented in Theorem 4.1 below.

Theorem 4.1 *Consider the system of Eq. 4.1 in closed-loop with x available at synchronous sampling time instants $\{t_{k\geq0}\}$ under the DMPC of Eqs. 4.10–4.21 based on a control law $u_1 = h(x)$ that satisfies the conditions of Eqs. 4.3–4.6. Let $\varepsilon_w > 0$, $\Delta > 0$ and $\rho > \rho_s > 0$ satisfy the following constraint:*

$$-\alpha_3\left(\alpha_2^{-1}(\rho_s)\right) + L'_x M\Delta + L'_w \theta \leq -\varepsilon_w/\Delta. \tag{4.24}$$

If $x(t_0) \in \Omega_\rho$, $\rho_{\min} \leq \rho$ and $N \geq 1$ where:

$$\rho_{\min} = \max\left\{V\left(x(t+\Delta)\right) : V\left(x(t)\right) \leq \rho_s\right\}, \tag{4.25}$$

then the state $x(t)$ of the closed-loop system is ultimately bounded in $\Omega_{\rho_{\min}}$.

Proof The proof consists of two parts. We first prove that the optimization problems of Eqs. 4.10–4.15 and 4.16–4.21 are feasible for all states $x \in \Omega_\rho$. Then we prove that, under the DMPC of Eqs. 4.10–4.21, the state of the system of Eq. 4.1 is ultimately bounded in a region that contains the origin.

    Part 1: We first prove the feasibility of LMPC 2 of Eqs. 4.10–4.15, and then the feasibility of LMPC 1 of Eqs. 4.16–4.21. All input trajectories of $u_2(t)$ such that $u_2(t) = 0$, $\forall t \in [t_k, t_{k+1})$ satisfy all the constraints (including the input constraint of Eq. 4.13 and the Lyapunov-based constraint of Eq. 4.15), thus the feasibility of LMPC 2 is guaranteed. The feasibility of LMPC 1 follows because all input trajectories $u_1(t)$ such that $u_1(t) = h(x(t_k))$, $\forall t \in [t_k, t_{k+1})$ are feasible solutions to the optimization problem of LMPC 1 since all such trajectories satisfy the input constraint of Eq. 4.18; this is guaranteed by the closed-loop stability property of the nonlinear control law $h(x)$ and the Lyapunov-based constraint of Eq. 4.21.

Part 2: From the conditions of Eqs. 4.3–4.6 and the constraints of Eqs. 4.15 and 4.21, if $x(t_k) \in \Omega_\rho$ it follows that:

$$\frac{\partial V(x(t_k))}{\partial x} f\left(x(t_k), u_1^*(t_k|t_k), u_2^*(t_k|t_k), 0\right)$$

$$\leq \frac{\partial V(x(t_k))}{\partial x} f\left(x(t_k), h(x(t_k)), u_2^*(t_k|t_k), 0\right)$$

$$\leq \frac{\partial V(x(t_k))}{\partial x} f\left(x(t_k), h(x(t_k)), 0, 0\right)$$

$$\leq -\alpha_3\left(\|x(t_k)\|\right). \qquad (4.26)$$

The time derivative of the Lyapunov function along the actual state trajectory $x(t)$ of the system of Eq. 4.1 in $t \in [t_k, t_{k+1})$ is given by:

$$\dot{V}\left(x(t)\right) = \frac{\partial V(x(t))}{\partial x} f\left(x(t), u_1^*(t_k|t_k), u_2^*(t_k|t_k), w(t)\right). \qquad (4.27)$$

Adding and subtracting $\frac{\partial V(x(t_k))}{\partial x} f(x(t_k), u_1^*(t_k|t_k), u_2^*(t_k|t_k), 0)$ and taking into account the conditions of Eq. 4.4, we obtain the following inequality:

$$\dot{V}\left(x(t)\right) \leq -\alpha_3\left(\|x(t_k)\|\right) + \frac{\partial V(x(t))}{\partial x} f\left(x(t), u_1^*(t_k|t_k), u_2^*(t_k|t_k), w(t)\right)$$

$$- \frac{\partial V(x(t_k))}{\partial x} f\left(x(t_k), u_1^*(t_k|t_k), u_2^*(t_k|t_k), 0\right). \qquad (4.28)$$

From the conditions of Eqs. 4.3–4.6, 4.9 and Eq. 4.28, the following inequality is obtained for all $x(t_k) \in \Omega_\rho/\Omega_{\rho_s}$:

$$\dot{V}\left(x(t)\right) \leq -\alpha_3\left(\alpha_2^{-1}(\rho_s)\right) + L_x'\|x(t) - x(t_k)\| + L_w'\|w\|. \qquad (4.29)$$

Taking into account Eq. 4.7 and the continuity of $x(t)$, the following bound can be written for all $t \in [t_k, t_{k+1})$:

$$\|x(t) - x(t_k)\| \leq M\Delta. \qquad (4.30)$$

Using this expression, we obtain the following bound on the time derivative of the Lyapunov function for $t \in [t_k, t_{k+1})$, for all initial states $x(t_k) \in \Omega_\rho/\Omega_{\rho_s}$:

$$\dot{V}\left(x(t)\right) \leq -\alpha_3\left(\alpha_2^{-1}(\rho_s)\right) + L_x'M\Delta + L_w'\theta. \qquad (4.31)$$

If the condition of Eq. 4.24 is satisfied, then there exists $\varepsilon_w > 0$ such that the following inequality holds for $x(t_k) \in \Omega_\rho/\Omega_{\rho_s}$:

$$\dot{V}\left(x(t)\right) \leq -\varepsilon_w/\Delta \qquad (4.32)$$

in $t \in [t_k, t_{k+1})$. Integrating this bound on $t \in [t_k, t_{k+1})$, we obtain that:

$$V\big(x(t_{k+1})\big) \leq V\big(x(t_k)\big) - \varepsilon_w, \tag{4.33}$$

$$V\big(x(t)\big) \leq V\big(x(t_k)\big), \quad \forall t \in [t_k, t_{k+1}) \tag{4.34}$$

for all $x(t_k) \in \Omega_\rho / \Omega_{\rho_s}$. Using Eqs. 4.33–4.34 recursively it is proved that, if $x(t_0) \in \Omega_\rho / \Omega_{\rho_s}$, the state converges to Ω_{ρ_s} in a finite number of sampling times without leaving the stability region. Once the state converges to $\Omega_{\rho_s} \subseteq \Omega_{\rho_{\min}}$, it remains inside $\Omega_{\rho_{\min}}$ for all times. This statement holds because of the definition of $\rho_{\min}$. This proves that the closed-loop system under the DMPC of Eqs. 4.10–4.21 is ultimately bounded in $\Omega_{\rho_{\min}}$. □

Remark 4.9 Referring to Theorem 4.1, the condition of Eq. 4.24 guarantees that if the state of the closed-loop system at a sampling time t_k is outside the level set $V(x(t_k)) = \rho_s$ but inside the level set $V(x(t_k)) = \rho$, the derivative of the Lyapunov function of the state of the closed-loop system is negative under the DMPC of Eqs. 4.10–4.21.

Remark 4.10 For continuous-time systems under continuous control implementation, a sufficient condition for set invariance is that the derivative of a Lyapunov function is negative on the boundary of a set. For systems with continuous-time dynamics and sample-and-hold control implementation, this condition is not sufficient because the derivative may become positive during the sampling period and the system may leave the set before a new sample is obtained. Referring to Theorem 4.1, $\rho_{\min}$ is the maximum value that the Lyapunov function can achieve in a time period of length Δ when $x(t_k) \in \Omega_{\rho_s}$. $\Omega_{\rho_{\min}}$ defines an invariant set for the state $x(t)$ under sample-and-hold implementation of the inputs of the DMPC of Eqs. 4.10–4.21.

Remark 4.11 To take advantage of both sets of manipulated inputs u_1 and u_2, one option is to design a centralized MPC. In order to guarantee robust stability of the closed-loop system, such a centralized MPC must include a set of stability constraints. To do this, we may use the LMPC of Eqs. 2.16–2.20 introduced in Chap. 2. This LMPC guarantees practical stability of the closed-loop system, allows for an explicit characterization of the stability region and yields a reduced complexity optimization problem. The LMPC for the system of Eq. 4.1 based on a nonlinear control law $h(x)$ satisfying the conditions of Eqs. 4.3–4.6 is based on the following optimization problem:

$$\min_{u_1, u_2 \in S(\Delta)} \int_{t_k}^{t_k+N} \Big[\big\|\tilde{x}(\tau)\big\|_{Q_c} + \big\|u_1(\tau)\big\|_{R_{c1}} + \big\|u_2(\tau)\big\|_{R_{c2}}\Big] d\tau, \tag{4.35}$$

$$\text{s.t.} \quad \dot{\tilde{x}}(t) = f\big(\tilde{x}(t), u_1(t), u_2(t), 0\big), \tag{4.36}$$

$$u_1(t) \in U_1, \tag{4.37}$$

$$u_2(t) \in U_2, \tag{4.38}$$

Fig. 4.2 Centralized LMPC
control architecture

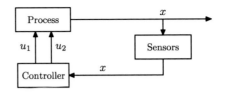

$$\tilde{x}(t_k) = x(t_k), \tag{4.39}$$

$$\frac{\partial V(x(t_k))}{\partial x} f\big(x(t_k), u_1(t_k), u_2(t_k), 0\big)$$

$$\leq \frac{\partial V(x(t_k))}{\partial x} f\big(x(t_k), h(x(t_k)), 0, 0\big), \tag{4.40}$$

where $\tilde{x}$ is the predicted trajectory of the nominal system for the input trajectory computed by this centralized LMPC.

The optimal solution to the optimization problem of Eqs. 4.35–4.40 is denoted by $u_{c1}^*(t|t_k)$ and $u_{c2}^*(t|t_k)$. The manipulated inputs of the closed-loop system under the above centralized LMPC are defined as follows

$$u_1(t) = u_{c1}^*(t|t_k), \quad \forall t \in [t_k, t_{k+1}), \tag{4.41}$$

$$u_2(t) = u_{c2}^*(t|t_k), \quad \forall t \in [t_k, t_{k+1}). \tag{4.42}$$

In what follows, we refer to this controller as the centralized LMPC. Figure 4.2 shows a schematic of this kind of control system.

Remark 4.12 The DMPC design presented in this section can be extended to include multiple MPCs using two different approaches. One approach is to use a one-directional sequential communication strategy (i.e., LMPC j sends information to LMPC $j-1$) and by letting each LMPC send along with its trajectory, all the trajectories received from previous controllers to its successor LMPC (i.e., LMPC j sends both its trajectory and the trajectories received from LMPC $j+1$ to LMPC $j-1$). A schematic of this approach is shown in Fig. 5.1 and it will be discussed in Chap. 5. Another approach is to have one master controller which communicates with all the other controllers using one-directional communication. This type of extension is shown in Fig. 5.14 which is a hierarchical type DMPC. More discussions of this type of DMPC and the corresponding approaches for handling communication disruptions in the DMPC can be found in Sect. 5.7.

4.4.3 Application to a Reactor–Separator Process

The example considered in this section is the reactor–separator process of Eqs. 3.52–3.63 described in Sects. 1.2.3 and 3.5.4 with the parameter values given

Table 4.1 Noise parameters

	σ_p	ϕ	θ_p
x_{A1}	1	0.7	0.25
x_{A2}	1	0.7	0.25
x_{A3}	1	0.7	0.25
x_{B1}	1	0.7	0.25
x_{B2}	1	0.7	0.25
x_{B3}	1	0.7	0.25
T_1	10	0.7	2.5
T_2	10	0.7	2.5
T_3	10	0.7	2.5

in Table 3.3. The manipulated inputs to the system are the heat inputs to the three vessels, Q_1, Q_2 and Q_3, and the feed stream flow rate to vessel 2, F_{20}.

The reactor–separator process of Eqs. 3.52–3.63 was numerically simulated using a standard Euler integration method. Process noise was added to the right-hand side of each ordinary differential equation in the process model to simulate disturbances/model uncertainty and it was generated as autocorrelated noise of the form $w_k = \phi w_{k-1} + \xi_k$ where $k = 0, 1, \ldots$ is the discrete time step of 0.001 h, ξ_k is generated by a normally distributed random variable with standard deviation σ_p, and ϕ is the autocorrelation factor and w_k is bounded by θ_p, that is $\|w_k\| \le \theta_p$. Table 4.1 contains the parameters used in generating the process noise.

We assume that the measurements of the temperatures T_1, T_2, T_3 and the measurements of the mass fractions x_{A1}, x_{B1}, x_{A2}, x_{B2}, x_{A3}, x_{B3} are available synchronously and continuously at time instants $\{t_{k\ge0}\}$ with $t_k = t_0 + k\Delta$, $k = 0, 1, \ldots$ where t_0 is the initial time and Δ is the sampling time. For the simulations carried out in this section, we pick the initial time to be $t_0 = 0$ and the sampling time to be $\Delta = 0.02$ h $= 1.2$ min.

The control objective is to regulate the system to a stable steady-state x_s corresponding to the operating point defined by Q_{1s}, Q_{2s}, Q_{3s} of u_{1s} and F_{20s} of u_{2s}. The steady-state values for u_{1s} and u_{2s} and the values of the steady-state are given in Table 4.2 and Table 4.3, respectively. Taking this control objective into account, the process model of Eqs. 3.52–3.63 belongs to the following class of nonlinear systems:

$$\dot{x}(t) = f\big(x(t)\big) + g_1\big(x(t)\big)u_1(t) + g_2\big(x(t)\big)u_2(t) + w(t), \qquad (4.43)$$

where $x^T = [x_1\ x_2\ x_3\ x_4\ x_5\ x_6\ x_7\ x_8\ x_9] = [x_{A1} - x_{A1s}\ x_{B1} - x_{B1s}\ T_1 - T_{1s}\ x_{A2} - x_{A2s}\ x_{B2} - x_{B2s}\ T_2 - T_{2s}\ x_{A3} - x_{A3s}\ x_{B3} - x_{B3s}\ T_3 - T_{3s}]$ is the state, $u_1^T = [u_{11}\ u_{12}\ u_{13}] = [Q_1 - Q_{1s}\ Q_2 - Q_{2s}\ Q_3 - Q_{3s}]$ and $u_2 = F_{20} - F_{20s}$ are the manipulated inputs which are subject to the constraints $|u_{1i}| \le 10^6$ KJ/h ($i = 1, 2, 3$) and $|u_2| \le 3$ m^3/h, and $w = w_k$ is a time varying bounded noise. The process of Eqs. 3.52–3.63 with the DMPC of Eqs. 4.10–4.21 is shown in Fig. 4.3.

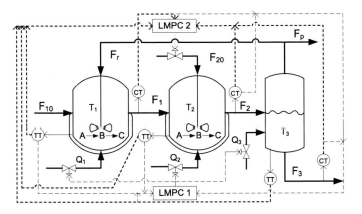

Fig. 4.3 Reactor–separator process with distributed control architecture

Table 4.2 Steady-state values for u_{1s} and u_{2s} of the reactor–separator process of Eqs. 3.52–3.63	Q_{1s}	12.6×10^5 [KJ/h]	Q_{3s}	11.88×10^5 [KJ/h]
	Q_{2s}	13.32×10^5 [KJ/h]	F_{20s}	5.04 [m³/h]

Table 4.3 Steady-state values for x_s of the reactor–separator process of Eqs. 3.52–3.63	x_{A1s}	0.605	x_{A2s}	0.605	x_{A3s}	0.346
	x_{B1s}	0.386	x_{B2s}	0.386	x_{B3s}	0.630
	T_{1s}	425.9 [K]	T_{2s}	422.6 [K]	T_{3s}	427.3 [K]

To illustrate the theoretical results, we first design the nonlinear control law $u_1 = h(x)$ which can stabilize the closed-loop system as follows [97]:

$$h(x) = \begin{cases} -\dfrac{L_f V + \sqrt{(L_f V)^2 + (L_{g1} V)^4}}{(L_{g1} V)^2} L_{g1} V & \text{if } L_{g1} V \neq 0, \\ 0 & \text{if } L_{g1} V = 0, \end{cases} \qquad (4.44)$$

where $L_f V = \frac{\partial V(x)}{\partial x} f(x)$ and $L_{g1} V = \frac{\partial V(x)}{\partial x} g_1(x)$ denote the Lie derivatives of the scalar function $V(x)$ with respect to the vector fields f and g_1, respectively. We consider a Lyapunov function $V(x) = x^T P x$ with P being the following weight matrix:

$$P = diag\left(5.2 \times 10^{12} [4\ 4\ 10^{-4}\ 4\ 4\ 10^{-4}\ 4\ 4\ 10^{-4}]\right). \qquad (4.45)$$

The values of the weights in P have been chosen in a way such that the control law of Eq. 4.44 stabilizes the closed-loop system globally (note that x_s is the only closed-loop system steady-state) and provides good closed-loop performance.

Based on the control law of Eq. 4.44, we design the centralized and the distributed LMPCs. In the simulations, the same parameters are used for both control designs. The prediction step is the same as the sampling time, that is $\Delta = 0.02\,\text{h} = 1.2\,\text{min}$; the prediction horizon is chosen to be $N = 6$; and the weight matrices for

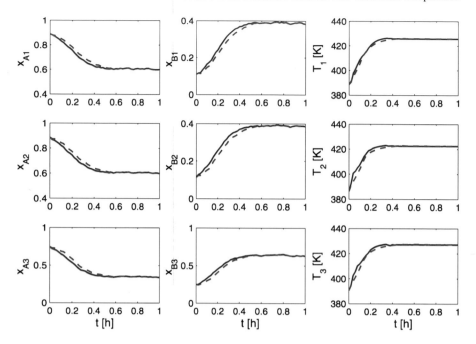

Fig. 4.4 State trajectories of the reactor–separator process of Eqs. 3.52–3.63 under the DMPC of Eqs. 4.10–4.21 (*solid lines*) and centralized LMPC of Eqs. 4.35–4.40 (*dashed lines*)

the LMPC designs are chosen as:

$$Q_c = diag\left(\left[2 \times 10^3\ 2 \times 10^3\ 2.5\ 2 \times 10^3\ 2 \times 10^3\ 2.5\ 2 \times 10^3\ 2 \times 10^3\ 2.5\right]\right),$$
$$R_{c1} = diag\left(\left[5 \times 10^{-12}\ 5 \times 10^{-12}\ 5 \times 10^{-12}\right]\right) \quad and \quad R_{c2} = 100.$$

$$(4.46)$$

The state and input trajectories of the process of Eqs. 3.52–3.63 under the DMPC of Eqs. 4.10–4.21 and the centralized LMPC of Eqs. 4.35–4.40 from the initial state:

$$x(0)^T = [0.890\ 0.110\ 388.7\ 0.886\ 0.113\ 386.3\ 0.748\ 0.251\ 390.6]. \qquad (4.47)$$

are shown in Figs. 4.4 and 4.5. Figure 4.4 shows that both the distributed and the centralized LMPC designs provide a similar closed-loop performance and drive the temperatures and the mass fractions in the closed-loop system close to the desired steady-state in about 0.3 h and 0.5 h, respectively.

We have also carried out a set of simulations to compare the DMPC of Eqs. 4.10–4.21 with the centralized LMPC of Eqs. 4.35–4.40 with the same parameters from a performance point of view. Table 4.4 shows the total cost computed for 15 different closed-loop simulations under the DMPC of Eqs. 4.10–4.21 and the centralized LMPC of Eqs. 4.35–4.40. To carry out this comparison, we have computed the total cost of each simulation with different operating conditions (different initial states

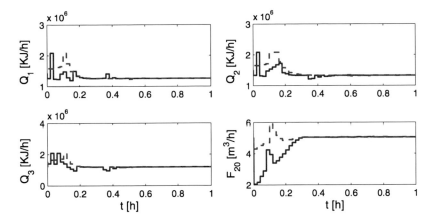

Fig. 4.5 Input trajectories of the reactor–separator process of Eqs. 3.52–3.63 under the DMPC of Eqs. 4.10–4.21 (*solid lines*) and centralized LMPC of Eqs. 4.35–4.40 (*dashed lines*)

and process noise) as follows:

$$\int_{t_0}^{t_M} \left[\left\| x(\tau) \right\|_{Q_c} + \left\| u_1(\tau) \right\|_{R_{c1}} + \left\| u_2(\tau) \right\|_{R_{c2}} \right] d\tau, \qquad (4.48)$$

where t_0 is the initial time of the simulations and $t_M = 1$ h is the end of the simulations. As we can see in Table 4.4, the DMPC of Eqs. 4.10–4.21 has a cost lower than the centralized LMPC of Eqs. 4.35–4.40 in 10 out of 15 simulations. This illustrates that in this example, the closed-loop performance of the DMPC of Eqs. 4.10–4.21 is comparable to the one of the centralized LMPC of Eqs. 4.35–4.40.

Remark 4.13 Table 4.4 shows that both controllers yield a similar performance for this particular process, but in general there is no guarantee that the total performance cost along the closed-loop system trajectories of either control scheme should be better than the other because the solution provided by the centralized LMPC of Eqs. 4.35–4.40 and the DMPC of Eqs. 4.10–4.21 are proved to be feasible and stabilizing but the convergence of the cost under DMPC of Eqs. 4.10–4.21 to the one under the centralized LMPC of Eqs. 4.35–4.40 is not established. This is because the communication between the two distributed MPCs is limited to one directional and moreover, the controllers are implemented in a receding horizon scheme and the prediction horizon is finite. In addition, there are disturbances modeled by stochastic noise in the simulations which introduce uncertainty in the results.

Moreover, we have studied the importance of communicating optimal input trajectories of LMPC 2 of Eqs. 4.10–4.15 to LMPC 1 of Eqs. 4.16–4.21. We have carried out a set of simulations in which both LMPC controllers operate in a decentralized manner; that is, LMPC 2 does not send its optimal input trajectory to LMPC 1 each sampling time (there is no communication between the two LMPCs). In order to evaluate its control input, LMPC 1 assumes that LMPC 2 applies the

Table 4.4 Total performance costs along the closed-loop trajectories of the reactor–separator process of Eqs. 3.52–3.63 under the DMPC of Eqs. 4.10–4.21 and the LMPC of Eqs. 4.35–4.40

sim.	DMPC of Eqs. 4.10–4.21	LMPC of Eqs. 4.35–4.40
1	65216	70868
2	70772	73112
3	57861	67723
4	62396	70914
5	60407	67109
6	83776	66637
7	61360	68897
8	47070	66818
9	79658	64342
10	65735	72819
11	62714	70951
12	76348	70547
13	49914	66869
14	89059	72431
15	78197	70257

steady-state input F_{20s}; that is $u_2 = 0$. The same parameters as in previous sets of simulations are used for the controllers. Figures 4.6 and 4.7 show the results under this decentralized LMPC scheme. From Fig. 4.6, we can see that for this particular example, this control scheme can not stabilize the system at the required steady-state. This result is expected because when there is no communication between the two distributed controllers, they can not coordinate their control actions and each controller views the input of the other controller as a disturbance that has to be rejected.

We have also carried out a set of simulations to compare the computation time needed to evaluate the distributed LMPCs (i.e., LMPC 1 of Eqs. 4.16–4.21 and LMPC 2 of Eqs. 4.10–4.15) with that of the centralized LMPC of Eqs. 4.35–4.40. The simulations have been carried out using MATLAB® in a PENTIUM® 3.20 GHz processor. The optimization problems have been solved using the built-in function *fmincon* of MATLAB®. To solve the ordinary differential equations in the process model, an Euler method with a fixed integration time of 0.001 h has been implemented in C programming language. For 50 evaluations, the mean time to solve the centralized LMPC of Eqs. 4.35–4.40 is 9.40 s; the mean times to solve LMPC 1 of Eqs. 4.16–4.21 and LMPC 2 of Eqs. 4.10–4.15 are 3.19 s and 4.53 s, respectively. From this set of simulations, we see that the computation time needed to solve the centralized LMPC of Eqs. 4.35–4.40 is larger than the sum of the values for LMPC 1 of Eqs. 4.16–4.21 and LMPC 2 of Eqs. 4.10–4.15 even though the closed-loop performance in terms of the total performance cost is comparable to the one of the DMPC of Eqs. 4.10–4.21. This is because the centralized LMPC of Eqs. 4.35–4.40 has to optimize both the inputs u_1 and u_2 in one optimization prob-

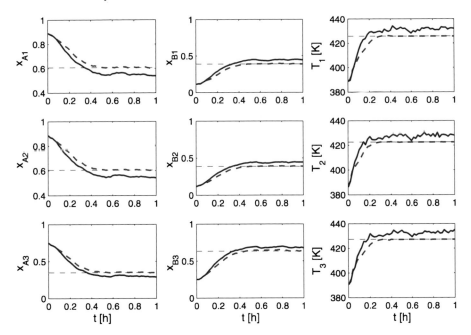

Fig. 4.6 State trajectories of the reactor–separator process of Eqs. 3.52–3.63 under the DMPC of Eqs. 4.10–4.21 without communication between the two LMPCs (*solid lines*) and with communication between the two LMPCs (*dashed lines*)

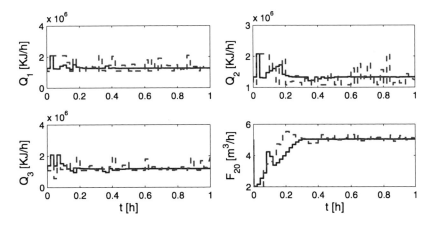

Fig. 4.7 Input trajectories of the reactor–separator process of Eqs. 3.52–3.63 under the DMPC of Eqs. 4.10–4.21 without communication between the two LMPCs (*solid lines*) and with communication between the two LMPCs (*dashed lines*)

lem and the DMPC of Eqs. 4.10–4.21 has to solve two smaller (in terms of decision variables) optimization problems.

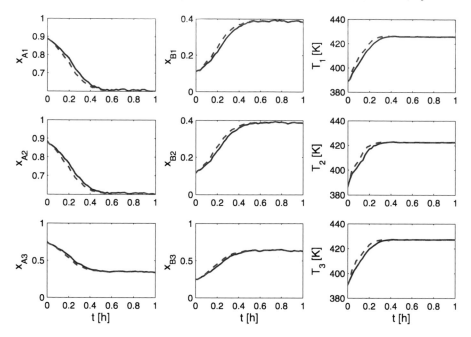

Fig. 4.8 State trajectories of the reactor–separator process of Eqs. 3.52–3.63 under the DMPC of Eqs. 4.10–4.21 with limited (*solid lines*) and unconstrained (*dashed lines*) evaluation time

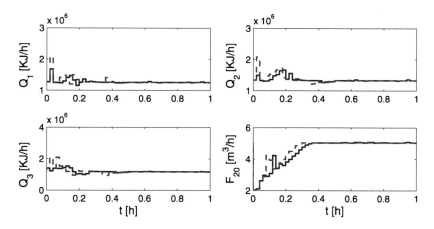

Fig. 4.9 Input trajectories of the reactor–separator process of Eqs. 3.52–3.63 under the DMPC of Eqs. 4.10–4.21 with limited (*solid lines*) and unconstrained (*dashed lines*) evaluation time

Following Remark 4.6, we have also carried out a set of simulations to illustrate that the optimization processes of LMPC 1 of Eqs. 4.16–4.21 and LMPC 2 of Eqs. 4.10–4.15 can be terminated at any time to get suboptimal solutions without loss of the closed-loop stability. In this set of simulations, we assume that the allowable evaluation times of LMPC 1 and LMPC 2 at each sampling time are 1 s and

2 s, and we terminate the two optimization processes when they have been carried out for 1 s and 2 s, respectively. The closed-loop state and input trajectories under the DMPC of Eqs. 4.10–4.21 with limited and unconstrained computation time are shown in Figs. 4.8 and 4.9. From Fig. 4.8, we see that the DMPC of Eqs. 4.10–4.21 with limited evaluation time can stabilize the closed-loop system but the state responses are slower, leading to a higher cost (57778) compared with the one (47117) obtained under the DMPC of Eqs. 4.10–4.21 with unconstrained computation time.

4.5   DMPC with Asynchronous Measurements

In this section, we design DMPC for the system of Eq. 4.1 subject to asynchronous measurements. In Sect. 4.6, we will extend the results to systems subject to delayed measurements.

4.5.1   Modeling of Asynchronous Measurements

We assume that the state of the system of Eq. 2.1, $x(t)$, is available asynchronously at time instants t_a where $\{t_{a \geq 0}\}$ is a random increasing sequence and the interval between two consecutive time instants is bounded by T_m; that is, the time sequence satisfies the condition of Eq. 2.22. This assumption is reasonable from a process control point of view.

4.5.2   DMPC Formulation

In Sect. 4.4, we introduced a DMPC design under the assumption of continuous, synchronous measurements. It was proved that the proposed control scheme guarantees practical stability of the closed-loop system and has the potential to maintain the closed-loop stability and performance in the face of new or failing controllers/actuators and to reduce computational burden in the evaluation of the optimal manipulated inputs compared with a centralized LMPC controller. However, when asynchronous measurements are present, the results obtained in Sect. 4.4 no longer hold. In order to simplify (but without loss of generality) the notations and description of the DMPC for system subject to asynchronous measurements (as well as asynchronous and delayed measurements discussed in Sect. 4.6), we will adopt the same strategy used in Sect. 4.4, that is, to design two LMPCs and coordinate their actions. The LMPC controllers computing the input trajectories of u_1 and u_2 are still referred to as LMPC 1 and LMPC 2, respectively. In this section, we extend the results of Sect. 4.4 to take into account asynchronous measurements explicitly, both in the constraints imposed on the LMPC designs and in the implementation

Fig. 4.10 DMPC design for systems subject to asynchronous measurements

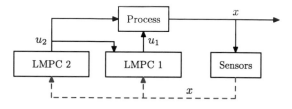

strategy. A schematic diagram of the considered closed-loop system is shown in Fig. 4.10.

In the presence of asynchronous measurements, the controllers need to operate in open-loop between successive state measurements. We take advantage of the MPC scheme to update the inputs based on a prediction obtained by the model. This is achieved by having the control actuators to store and implement the last computed optimal input trajectories. The implementation strategy is as follows:

1. When a measurement is available at t_a, LMPC 2 computes the optimal input trajectory of u_2.
2. LMPC 2 sends the entire optimal input trajectory to its actuators and also sends the entire optimal input trajectory to LMPC 1.
3. Once LMPC 1 receives the entire optimal input trajectory for u_2, it evaluates the optimal input trajectory of u_1.
4. LMPC 1 sends the entire optimal input trajectory to its actuators.
5. When a new measurement is received ($a \leftarrow a + 1$), go to Step 1.

We first design the optimization problem of LMPC 2 for systems subject to asynchronous measurements. This optimization problem depends on the latest state measurement $x(t_a)$. In order to make a decision, LMPC 2 must assume LMPC 1 applies the nonlinear control law $u_1 = h(x)$. The LMPC 2 is based on the following optimization problem:

$$\min_{u_2 \in S(\Delta)} \int_{t_a}^{t_a + N\Delta} \left[\left\| \tilde{x}(\tau) \right\|_{Q_c} + \left\| u_1(\tau) \right\|_{R_{c1}} + \left\| u_2(\tau) \right\|_{R_{c2}} \right] d\tau, \tag{4.49}$$

$$\text{s.t.} \quad \dot{\tilde{x}}(t) = f\big(\tilde{x}(t), u_1(t), u_2(t), 0\big), \tag{4.50}$$

$$u_1(t) = h\big(\tilde{x}(t_a + j\Delta)\big), \quad \forall t \in \big[t_a + j\Delta, t_a + (j+1)\Delta\big), \tag{4.51}$$

$$u_2(t) \in U_2, \tag{4.52}$$

$$\dot{\hat{x}}(t) = f\big(\hat{x}(t), h\big(\hat{x}(t_a + j\Delta)\big), 0, 0\big),$$
$$\forall t \in \big[t_a + j\Delta, t_a + (j+1)\Delta\big), \tag{4.53}$$

$$\tilde{x}(t_a) = \hat{x}(t_a) = x(t_a), \tag{4.54}$$

$$V\big(\tilde{x}(t)\big) \leq V\big(\hat{x}(t)\big), \quad \forall t \in [t_a, t_a + N_R\Delta), \tag{4.55}$$

where $\tilde{x}$ is the predicted trajectory of the nominal system with u_2 being the input trajectory computed by the LMPC of Eqs. 4.49–4.55 (i.e., LMPC 2) and u_1

being the nonlinear control law $h(x)$ applied in a sample-and-hold fashion with
$j = 0, \ldots, N - 1$, $\hat{x}$ is the predicted trajectory of the nominal system with u_1 being $h(x)$ applied in a sample-and-hold fashion and $u_2 = 0$, and N_R is the smallest
integer that satisfies the inequality $T_m \leq N_R \Delta$. To take full advantage of the nominal model in the computation of the control action, we take $N \geq N_R$. The optimal
solution to this optimization problem is denoted by $u_{a2}^*(t|t_a)$ which is defined for
$t \in [t_a, t_a + N\Delta)$. Once the optimal input trajectory of u_2 is available, it is sent to
LMPC 1 as well as to its corresponding control actuators.

Note that the constraints of Eqs. 4.53–4.54 generate a reference state trajectory
(i.e., a reference Lyapunov function trajectory) of the closed-loop system; and the
constraint of Eq. 4.55 ensures that the predicted decrease of the Lyapunov function
from t_a to $t_a + N_R\Delta$, if $u_1 = h(x)$ and $u_2 = u_{a2}^*(t|t_a)$ are applied, is at least equal
to the one obtained from the constraint of Eq. 4.53. By imposing the constraint of
Eq. 4.55 (as well as the constraint of Eq. 4.62), we can prove that the distributed control system inherits the stability properties of the nonlinear control law $h(x)$ when
it is implemented in a sample-and-hold fashion. Note also that we have considered
input constraints (see Eq. 4.52).

The optimization problem of LMPC 1 for systems subject to asynchronous measurements depends on $x(t_a)$ and the decision taken by LMPC 2 of Eqs. 4.49–4.55
(i.e., $u_{a2}^*(t|t_a)$). This allows LMPC 1 to compute a u_1 such that the closed-loop per-
formance is optimized, while guaranteeing that the stability properties of the nonlinear control law $h(x)$ are preserved. Specifically, LMPC 1 is based on the following
optimization problem:

$$\min_{u_1 \in S(\Delta)} \int_{t_a}^{t_a+N\Delta} \left[\|\check{x}(\tau)\|_{Q_c} + \|u_1(\tau)\|_{R_{c1}} + \|u_2(\tau)\|_{R_{c2}} \right] d\tau, \tag{4.56}$$

$$\text{s.t.} \quad \dot{\check{x}}(t) = f\big(\check{x}(t), u_1(t), u_2(t), 0\big), \tag{4.57}$$

$$\dot{\tilde{x}}(t) = f\big(\tilde{x}(t), h\big(\tilde{x}(t_a + j\Delta)\big), u_2(t), 0\big),$$
$$\forall t \in \big[t_a + j\Delta, t_a + (j+1)\Delta\big), \tag{4.58}$$

$$u_2(t) = u_{a2}^*(t|t_a), \tag{4.59}$$

$$u_1(t) \in U_1, \tag{4.60}$$

$$\check{x}(t_a) = \tilde{x}(t_a) = x(t_a), \tag{4.61}$$

$$V\big(\check{x}(t)\big) \leq V\big(\tilde{x}(t)\big), \quad \forall t \in [t_a, t_a + N_R\Delta), \tag{4.62}$$

where $\check{x}$ is the predicted trajectory of the nominal system if $u_2 = u_{a2}^*(t|t_a)$ and u_1
computed by the LMPC 1 of Eqs. 4.56–4.62 are applied, and $\tilde{x}$ is the predicted
trajectory of the nominal system if $u_2 = u_{a2}^*(t|t_a)$ and the nonlinear control law
$h(x)$ are applied in a sample-and-hold fashion with $j = 0, \ldots, N - 1$. The optimal
solution to this optimization problem is denoted by $u_{a1}^*(t|t_a)$ which is defined for
$t \in [t_a, t_a + N\Delta)$. The constraint of Eq. 4.62 guarantees that the predicted decrease
of the Lyapunov function from t_a to $t_a + N_R\Delta$, if $u_1 = u_{a1}^*(t|t_a)$ and $u_2 = u_{a2}^*(t|t_a)$

are applied, is at least equal to the one obtained when $u_1 = h(x)$ and $u_2 = u_{a2}^*(t|t_a)$ are applied. Note that the trajectory $\tilde{x}(t)$ predicted by the constraint of Eq. 4.58 is the same as the optimal trajectory predicted by LMPC 2 of Eqs. 4.49–4.55. This trajectory will be used in the proof of the closed-loop stability properties of the controller. The manipulated inputs of the distributed control scheme of Eqs. 4.49–4.62 are defined as follows:

$$u_1(t) = u_{a1}^*(t|t_a), \quad \forall t \in [t_a, t_{a+1}), \tag{4.63}$$

$$u_2(t) = u_{a2}^*(t|t_a), \quad \forall t \in [t_a, t_{a+1}). \tag{4.64}$$

Note that, as explained before, the actuators apply the last evaluated optimal input trajectories between two successive state measurements.

4.5.3 Stability Properties

In this subsection, we prove that the distributed control scheme of Eqs. 4.49–4.62 inherits the stability properties of the nonlinear control law $h(x)$ implemented in a sample-and-hold fashion. This property is presented in Theorem 4.2 below.

Theorem 4.2 *Consider the system of Eq. 4.1 in closed-loop with x available at asynchronous sampling time instants $\{t_{a \geq 0}\}$, satisfying the condition of Eq. 2.22, under the DMPC of Eqs. 4.49–4.62 based on a control law $u_1 = h(x)$ that satisfies the conditions of Eqs. 4.3–4.6. Let $\Delta, \varepsilon_s > 0$, $\rho > \rho_{\min} > 0$, $\rho > \rho_s > 0$ and $N \geq N_R \geq 1$ satisfy the condition of Eq. 2.31 and the following inequality:*

$$-N_R \varepsilon_s + f_V\big(f_W(N_R \Delta)\big) < 0 \tag{4.65}$$

with $f_V(\cdot)$ and $f_W(\cdot)$ defined in Eqs. 2.49 and 2.43, respectively, and N_R being the smallest integer satisfying $N_R \Delta \geq T_m$. If $x(t_0) \in \Omega_\rho$, then $x(t)$ is ultimately bounded in $\Omega_{\rho_a} \subseteq \Omega_\rho$ where:

$$\rho_a = \rho_{\min} + f_V\big(f_W(N_R \Delta)\big) \tag{4.66}$$

with $\rho_{\min}$ defined in Eq. 4.25.

Proof In order to prove that the closed-loop system is ultimately bounded in a region that contains the origin, we prove that $V(x(t_a))$ is a decreasing sequence of values with a lower bound.

  Part 1: In this part, we prove that the stability results stated in Theorem 4.2 hold in the case that $t_{a+1} - t_a = T_m$ for all a and $T_m = N_R \Delta$. This case corresponds to the worst possible situation in the sense that LMPC 1 of Eqs. 4.56–4.62 and LMPC 2 of Eqs. 4.49–4.55 need to operate in open-loop for the maximum possible amount of time. In order to simplify the notation, we assume that all the variables used in this proof refer to the different optimization variables of the problems solved at time step t_a; that is, $\hat{x}(t_{a+1})$ is obtained from the nominal closed-loop trajectory of the system of Eq. 4.1 under the Lyapunov-based controller $u_1 = h(x)$ implemented in a sample-and-hold fashion and $u_2 = 0$ starting from $x(t_a)$. By Proposition 2.1 and the fact that $t_{a+1} = t_a + N_R \Delta$, the following inequality can be obtained:

$$V\big(\hat{x}(t_{a+1})\big) \le \max\big\{V\big(\hat{x}(t_a)\big) - N_R \varepsilon_s, \rho_{\min}\big\}. \tag{4.67}$$

From the constraints of Eqs. 4.55 and 4.62 in LMPC 2 and LMPC 1, the following inequality can be written:

$$V\big(\check{x}(t)\big) \le V\big(\tilde{x}(t)\big) \le V\big(\hat{x}(t)\big), \quad \forall t \in [t_a, t_a + N_R \Delta). \tag{4.68}$$

From inequalities of Eqs. 4.67 and 4.68 and taking into account that $\hat{x}(t_a) = \tilde{x}(t_a) = \check{x}(t_a) = x(t_a)$, the following inequality is obtained:

$$V\big(\check{x}(t_{a+1})\big) \le \max\big\{V\big(x(t_a)\big) - N_R \varepsilon_s, \rho_{\min}\big\}. \tag{4.69}$$

When $x(t) \in \Omega_\rho$ for all times (this point will be proved below), we can apply Proposition 2.3 to obtain the following inequality:

$$V\big(x(t_{a+1})\big) \le V\big(\check{x}(t_{a+1})\big) + f_V\big(\|\check{x}(t_{a+1}) - x(t_{a+1})\|\big). \tag{4.70}$$

Applying Proposition 2.2, we obtain the following upper bound on the deviation of $\check{x}(t)$ from $x(t)$:

$$\|x(t_{k+1}) - \check{x}(t_{k+1})\| \le f_W(N_R \Delta). \tag{4.71}$$

From the inequalities of Eqs. 4.70 and 4.71, the following upper bound on $V(x(t_{k+1}))$ can be written:

$$V\big(x(t_{a+1})\big) \le V\big(\check{x}(t_{a+1})\big) + f_V\big(f_W(N_R \Delta)\big). \tag{4.72}$$

Using the inequality of Eq. 4.69, we can rewrite the inequality of Eq. 4.72 as follows:

$$V\big(x(t_{a+1})\big) \le \max\big\{V\big(x(t_a)\big) - N_R \varepsilon_s, \rho_{\min}\big\} + f_V\big(f_W(N_R \Delta)\big). \tag{4.73}$$

If the condition of Eq. 4.65 is satisfied, from the inequality of Eq. 4.73, we know that there exists $\varepsilon_w > 0$ such that the following inequality holds:

$$V\big(x(t_{a+1})\big) \le \max\big\{V\big(x(t_a)\big) - \varepsilon_w, \rho_a\big\}, \tag{4.74}$$

which implies that if $x(t_a) \in \Omega_\rho/\Omega_{\rho_a}$, then $V(x(t_{a+1})) < V(x(t_a))$, and if $x(t_a) \in \Omega_{\rho_a}$, then $V(x(t_{a+1})) \le \rho_a$.

Because the upper bound on the difference between the Lyapunov function of the actual trajectory x and the nominal trajectory $\check{x}$ is a strictly increasing function of time (see Proposition 2.2 and Proposition 2.3 for the expressions of $f_W(\cdot)$ and $f_V(\cdot)$), the inequality of Eq. 4.74 also implies that:

$$V\big(x(t)\big) \le \max\big\{V\big(x(t_a)\big), \rho_a\big\}, \quad \forall t \in [t_a, t_{a+1}). \tag{4.75}$$

Using the inequality of Eq. 4.75 recursively, it can be proved that if $x(t_0) \in \Omega_\rho$, then the closed-loop trajectories of the system of Eq. 4.1 under the DMPC of Eqs. 4.49–4.62 stay in Ω_ρ for all times (i.e., $x(t) \in \Omega_\rho$, $\forall t$). Moreover, using the inequality of Eq. 4.74 recursively, it can be proved that if $x(t_0) \in \Omega_\rho$, the closed-loop trajectories of the system of Eq. 4.1 under the DMPC of Eqs. 4.49–4.62 satisfy:

$$\limsup_{t \to \infty} V\big(x(t)\big) \leq \rho_a. \tag{4.76}$$

This proves that $x(t) \in \Omega_\rho$ for all times and $x(t)$ is ultimately bounded in Ω_{ρ_a} for the case when $t_{a+1} - t_a = T_m$ for all a and $T_m = N_R \Delta$.

Part 2: In this part, we extend the results proved in Part 1 to the general case, that is, $t_{a+1} - t_a \leq T_m$ for all a and $T_m \leq N_R \Delta$ which implies that $t_{a+1} - t_a \leq N_R \Delta$. Because $f_V(\cdot)$ and $f_W(\cdot)$ are strictly increasing functions of their arguments and $f_V(\cdot)$ is convex, following similar steps as in Part 1, it can be shown that the inequality of Eq. 4.73 still holds. This proves that the stability results stated in Theorem 4.2 hold. ☐

4.5.4 Application to a Reactor–Separator Process

Consider the reactor–separator process of Eqs. 3.52–3.63 described in Sect. 1.2.3 with the parameter values given in Table 3.3. As in the simulations carried out in Sect. 4.4.3, in this section, the process was also numerically simulated using a standard Euler integration method, and bounded process noise was added to all the simulations to simulate disturbances/model uncertainty. The manipulated inputs to the system are the heat inputs, Q_1, Q_2 and Q_3, and the feed stream flow rate to vessel 2, F_{20}. For each set of steady-state inputs Q_{1s}, Q_{2s}, Q_{3s} and F_{20s} corresponding to a different operating condition, the process has one stable steady-state x_s. The control objective is to steer the process from the initial state:

$$x_0^T = [0.89\ 0.11\ 388.7\ 0.11\ 386.3\ 0.75\ 0.25\ 390.6], \tag{4.77}$$

to the steady-state:

$$x_s^T = [0.61\ 0.39\ 425.9\ 0.61\ 0.39\ 422.6\ 0.35\ 0.63\ 427.3], \tag{4.78}$$

which is the steady-state corresponding to the operating condition: $Q_{1s} = 12.6 \times 10^5$ KJ/h, $Q_{3s} = 11.88 \times 10^5$ KJ/h, $Q_{2s} = 13.32 \times 10^5$ KJ/h and $F_{20s} = 5.04$ m^3/h. The process belongs to the following class of nonlinear systems:

$$\dot{x}(t) = f\big(x(t)\big) + g_1\big(x(t)\big)u_1(t) + g_2\big(x(t)\big)u_2(t) + w(t), \tag{4.79}$$

where $x^T = [x_1\ x_2\ x_3\ x_4\ x_5\ x_6\ x_7\ x_8\ x_9] = [x_{A1} - x_{A1s}\ x_{B1} - x_{B1s}\ T_1 - T_{1s}\ x_{A2} - x_{A2s}\ x_{B2} - x_{B2s}\ T_2 - T_{2s}\ x_{A3} - x_{A3s}\ x_{B3} - x_{B3s}\ T_3 - T_{3s}]$ is the state, $u_1^T =$

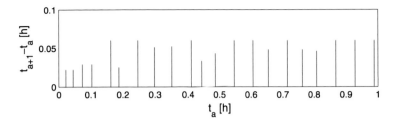

Fig. 4.11 Asynchronous measurement sampling times $\{t_{a\geq0}\}$ with $T_m = 3\Delta$: the x-axis indicates $\{t_{a\geq0}\}$ and the y-axis indicates the size of the interval between t_a and t_{a-1}

$[u_{11}\ u_{12}\ u_{13}] = [Q_1 - Q_{1s}\ Q_2 - Q_{2s}\ Q_3 - Q_{3s}]$ and $u_2 = F_{20} - F_{20s}$ are the manipulated inputs which are subject to the constraints $|u_{1i}| \leq 10^6$ KJ/h ($i = 1, 2, 3$) and $|u_2| \leq 3$ m³/h, and w is a bounded noise.

We use the same design of $h(x)$ as in Eq. 4.44, and we consider the same Lyapunov function $V(x) = x^T P x$ with $P = diag(5.2 \times 10^{12}[4\ 4\ 10^{-4}\ 4\ 4\ 10^{-4}\ 4\ 4\ 10^{-4}])$ as in Sect. 4.4.3.

For the simulations carried out in this section, it is assumed that the state measurements of the process are available asynchronously at time instants $\{t_{a\geq0}\}$ with an upper bound $T_m = 3\Delta$ on the maximum interval between two successive asynchronous state measurements, where Δ is the controller and sensor sampling time and is chosen to be $\Delta = 0.02$ h $= 1.2$ min. Based on the Lyapunov-based controller $h(x)$, we design LMPC 1 and LMPC 2. The prediction horizons of both LMPC 1 of Eqs. 4.56–4.62 and LMPC 2 of Eqs. 4.49–4.55 are chosen to be $N = 6$ and N_R is chosen to be 3 so that $N_R\Delta \geq T_m$. The weighting matrices for the LMPCs are chosen in a way such that the DMPC of Eqs. 4.10–4.21 presented in Sect. 4.4 and the DMPC of Eqs. 4.49–4.62 can both stabilize the closed-loop system with state measurements obtained at each sampling time. Specifically, the weighting matrices are chosen as follows:

$$Q_c = diag\left(10^3[2\ 2\ 0.0025\ 2\ 2\ 0.0025\ 2\ 2\ 0.0025]\right), \tag{4.80}$$

$R_{c1} = diag([5 \times 10^{-12}\ 5 \times 10^{-12}\ 5 \times 10^{-12}])$ and $R_{c2} = 100$.

To model the time sequence $\{t_{a\geq0}\}$, we use an upper bounded random Poisson process. The Poisson process is defined by the number of events per unit time W. The interval between two successive concentration sampling times (events of the Poisson process) is given by $\Delta_a = \min\{-\ln\chi / W, T_m\}$, where χ is a random variable with uniform probability distribution between 0 and 1. This generation ensures that $\max_a\{t_{a+1} - t_a\} \leq T_m$. In this example, W is chosen to be $W = 20$. The generated time sequence $\{t_{a\geq0}\}$ for a simulation length of 1.0 h is shown in Fig. 4.11 and the average time interval between two successive time instants is 0.046 h.

In this set of simulations, when the system operates in open-loop, all the control designs to be tested use their last evaluated optimal input trajectories. The state and input trajectories of the process of Eqs. 3.52–3.63 in closed-loop under the DMPC of Eqs. 4.49–4.62 taking into account asynchronous measurements explicitly and

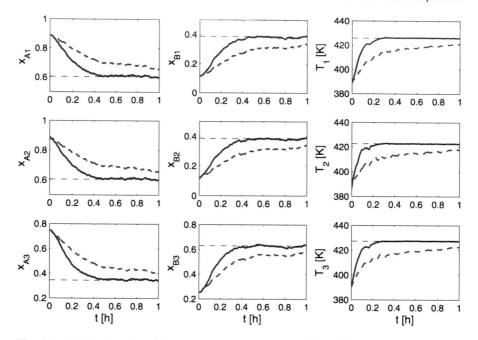

Fig. 4.12 State trajectories of the reactor–separator process of Eqs. 3.52–3.63 under the DMPC of Eqs. 4.49–4.62 (*solid lines*) and the DMPC of Eqs. 4.10–4.21 (*dashed lines*) in the presence of asynchronous measurements

the DMPC of Eqs. 4.10–4.21 are shown in Figs. 4.12 and 4.13. In Fig. 4.12, it can be seen that the DMPC of Eqs. 4.49–4.62 provides a better performance and is able to stabilize the process at the desired steady state in about 0.5 h; the DMPC of Eqs. 4.10–4.21 fails to drive the state of the process to the desired steady state within 1 h because it does not account for the asynchronous measurements.

4.6 DMPC with Delayed Measurements

In this section, we consider DMPC of systems subject to asynchronous measurements involving time-varying delays.

4.6.1 Modeling of Delayed Measurements

We assume that the state of the system of Eq. 4.1 is received by the controllers at asynchronous time instants t_a where $\{t_{a\geq0}\}$ is a random increasing sequence of times and that there exists an upper bound T_m on the interval between two successive measurements. In order to model delays in measurements, another auxiliary variable d_a

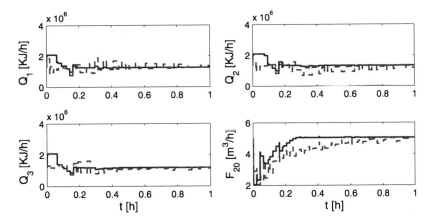

Fig. 4.13 Input trajectories of the reactor–separator process of Eqs. 3.52–3.63 under the DMPC of Eqs. 4.49–4.62 (*solid lines*) and the DMPC of Eqs. 4.10–4.21 (*dashed lines*) in the presence of asynchronous measurements

Fig. 4.14 DMPC design for systems subject to delayed measurements

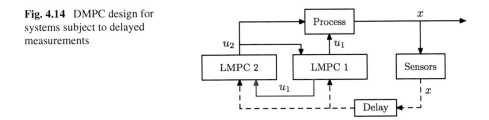

is introduced to indicate the delay corresponding to the measurement received at time t_a, that is, at time t_a, the measurement $x(t_a - d_a)$ is received. We assume that the delays associated with the measurements are smaller than an upper bound D. As explained in Sect. 2.8.1, the maximum amount of time the system might operate in open-loop following t_a is $D + T_m - d_a$; please also see Fig. 2.6 for a possible sequence of delayed measurements. This upper bound will be used in the formulation of the DMPC design for systems subject to delayed measurements below.

4.6.2 DMPC Formulation

As in Sects. 2.8 and 3.6, we take advantage of the system model both to estimate the current system state from a delayed measurement and to control the system in open-loop when new information is not available. Specifically, when a delayed measurement is received, the controllers use the system model and the manipulated inputs that have been applied to the system to get an estimate of the current state and then an MPC optimization problem is solved in order to decide the optimal future input trajectory that will be applied until new measurements are received. However, in the distributed schemes previously presented (see Fig. 4.10), LMPC 2 does not know the input trajectory which has been implemented by LMPC 1 because there is only one-directional communication from LMPC 2 to LMPC 1. In order to get a good estimate of the current state from a delayed measurement, the DMPC structure shown in Fig. 4.10 needs to be modified to have bidirectional communication so that LMPC 1 can send its optimal input trajectory to LMPC 2. A schematic of the DMPC scheme for systems subject to asynchronous and delayed measurements considered in this section is shown in Fig. 4.14. When at t_a, a delayed measurement $x(t_a - d_a)$ is received, the information sent from LMPC 1 to LMPC 2 allows LMPC 2 to estimate the current state by using the system model of Eq. 4.1 and the input trajectories $u_1(t)$ (which has received from LMPC 1) and $u_2(t)$ (which LMPC 2 has stored in memory) applied in $t \in [t_a - d_a, t_a)$. The implementation strategy in the presence of delayed measurements is as follows:

1. When a measurement $x(t_a - d_a)$ is available at t_a, LMPC 2 checks whether the measurement provides new information. If $t_a - d_a > \max_{l<a} t_l - d_l$, go to Step 2. Else the measurement does not contain new information and is discarded, go to Step 6.
2. LMPC 2 estimates the current state of the system $\tilde{x}(t_a)$ and computes the optimal input trajectory of u_2 based on $\tilde{x}(t_a)$.
3. LMPC 2 sends its entire optimal input trajectory to its actuators and also sends $\tilde{x}(t_a)$ and its entire optimal input trajectory to LMPC 1.
4. Once LMPC 1 receives $\tilde{x}(t_a)$ and the entire optimal input trajectory for u_2, it evaluates the optimal input trajectory of u_1 based on $\tilde{x}(t_a)$.
5. LMPC 1 sends its entire optimal input trajectory to its actuators and LMPC 2.
6. When a new measurement is received ($a \leftarrow a + 1$), go to Step 1.

The LMPC 2 for systems subject to delayed measurements is based on the following optimization problem:

$$\min_{u_2 \in S(\Delta)} \int_{t_a}^{t_a+N\Delta} \left[\|\tilde{x}(\tau)\|_{Q_c} + \|u_1(\tau)\|_{R_{c1}} + \|u_2(\tau)\|_{R_{c2}} \right] d\tau, \tag{4.81}$$

$$\text{s.t.} \quad \dot{\tilde{x}}(t) = f\left(\tilde{x}(t), u_1(t), u_2(t), 0\right), \quad \forall t \in [t_a - d_a, t_a + N\Delta), \tag{4.82}$$

$$u_1(t) = u_{d1}^*(t), \quad \forall t \in [t_a - d_a, t_a), \tag{4.83}$$

$$u_2(t) = u_{d2}^*(t), \quad \forall t \in [t_a - d_a, t_a), \tag{4.84}$$

$$u_1(t) = h\left(\tilde{x}(t_a + j\Delta)\right), \quad \forall t \in \left[t_a + j\Delta, t_a + (j+1)\Delta\right), \tag{4.85}$$

$$u_2(t) \in U_2, \tag{4.86}$$

$$\tilde{x}(t_a - d_a) = x(t_a - d_a), \tag{4.87}$$

$$\dot{\hat{x}}(t) = f\left(\hat{x}(t), h\left(\hat{x}(t_a + j\Delta)\right), 0, 0\right),$$

$$\forall t \in \left[t_a + j\Delta, t_a + (j+1)\Delta\right), \tag{4.88}$$

$$\hat{x}(t_a) = \tilde{x}(t_a), \tag{4.89}$$

$$V\big(\tilde{x}(t)\big) \le V\big(\hat{x}(t)\big), \quad \forall t \in [t_a, t_a + N_{D,a}\Delta), \tag{4.90}$$

where $j = 0, \ldots, N-1$, and $N_{D,a}$ is the smallest integer satisfying $N_{D,a}\Delta \ge T_m + D - d_a$ and $u_{d1}^*(t), u_{d2}^*(t)$ are the latest input trajectories sent by the controllers to the actuators. The optimal solution to this optimization problem is denoted by $u_{d2}^*(t|t_a)$ which is defined for $t \in [t_a, t_a + N\Delta)$. Once this optimal input trajectory of u_2 is available, it is sent to the control actuators controlled by LMPC 2 and to LMPC 1 together with the estimate of the current state $\tilde{x}(t_a)$.

There are two types of calculations in the optimization problem of Eqs. 4.81–4.90. The first type of calculation is to estimate the current state $\tilde{x}(t_a)$ based on the delayed measurement $x(t_a - d_a)$ and input values that have applied to the system from $t_a - d_a$ to t_a (the constraints of Eqs. 4.82, 4.83, 4.84 and 4.87). The second type of calculation is to evaluate the optimal input trajectory of u_2 based on $\tilde{x}(t_a)$ while satisfying the input constraint of Eq. 4.86 and the constraint of Eq. 4.90. The constraint of Eq. 4.90 is required to ensure the practical closed-loop stability. Note that the length of the constraint $N_{D,a}$ depends on the current delay d_a, so it may have different values at different time instants and has to be updated before solving the optimization problem of Eqs. 4.81–4.90.

The LMPC 1 for systems subject to delayed measurements depends on $\tilde{x}(t_a)$ and $u_{d2}^*(t|t_a)$. Specifically, it is based on the following optimization problem:

$$\min_{u_1 \in S(\Delta)} \int_{t_a}^{t_a+N\Delta} \Big[\|\check{x}(\tau)\|_{Q_c} + \|u_1(\tau)\|_{R_{c1}} + \|u_2(\tau)\|_{R_{c2}} \Big] d\tau, \tag{4.91}$$

$$\text{s.t.} \quad \dot{\tilde{x}}(t) = f\big(\tilde{x}(t), h\big(\tilde{x}(t_a + j\Delta)\big), u_2(t), 0\big),$$

$$\forall t \in \big[t_a + j\Delta, t_a + (j+1)\Delta\big), \tag{4.92}$$

$$\dot{\check{x}}(t) = f\big(\check{x}(t), u_1(t), u_2(t), 0\big), \tag{4.93}$$

$$u_2(t) = u_{d2}^*(t|t_a), \tag{4.94}$$

$$u_1(t) \in U_1, \tag{4.95}$$

$$\check{x}(t_a) = \tilde{x}(t_a), \tag{4.96}$$

$$V\big(\check{x}(t)\big) \le V\big(\tilde{x}(t)\big), \quad \forall t \in [t_a, t_a + N_{D,a}\Delta). \tag{4.97}$$

The optimal solution to the optimization problem of Eqs. 4.91–4.97 is denoted as $u_{d2}^*(t|t_a)$ which is defined for $t \in [t_a, t_a + N\Delta)$ and it is sent to the control actuators controlled by LMPC 1 and LMPC 2. Note that LMPC 1 gets $\tilde{x}(t_a)$ from LMPC 2, so it does not need to estimate the current state and only needs to evaluate the optimal input trajectory of u_1 based on $\tilde{x}(t_a)$ while satisfying the input constraint of Eq. 4.95 and the constraint of Eq. 4.97. The constraint of Eq. 4.97 is required to ensure closed-loop practical stability.

The manipulated inputs of the DMPC of Eq. 4.81–4.97 for systems subject to asynchronous and delayed measurements are defined as follows:

$$u_1(t) = u_{d1}^*(t|t_a), \quad \forall t \in [t_a, t_{a+i}), \tag{4.98}$$

$$u_2(t) = u_{d2}^*(t|t_a), \quad \forall t \in [t_a, t_{k+i}) \tag{4.99}$$

for all t_a such that $t_a - d_a > \max_{l<a} t_l - d_l$ and for a given t_a, the variable i denotes the smallest integer that satisfies $t_{a+i} - d_{a+i} > t_a - d_a$.

4.6.3 Stability Properties

In this subsection, we present the stability property of the distributed control scheme of Eqs. 4.81–4.97. This property is presented in Theorem 4.3 below.

Theorem 4.3 *Consider the system of Eq. 4.1 in closed-loop with x available at asynchronous sampling time instants $\{t_{a\geq 0}\}$ involving time-varying delays such that $d_a \leq D$ for all $a \geq 0$, satisfying the condition of Eq. 2.22, under the DMPC of Eqs. 4.81–4.97 based on a control law $u_1 = h(x)$ that satisfies the conditions of Eqs. 4.3–4.6. Let $\Delta, \varepsilon_s > 0, \rho > \rho_{\min} > 0, \rho > \rho_s > 0, N \geq 1$ and $D \geq 0$ satisfy the condition of Eq. 2.31 and the following inequality:*

$$-N_R \varepsilon_s + f_V\big(f_W(N_D \Delta)\big) + f_V\big(f_W(D)\big) < 0 \tag{4.100}$$

with $f_V(\cdot)$ and $f_W(\cdot)$ defined in Eqs. 2.49 and 2.43, respectively, N_D the smallest integer satisfying $N_D \Delta \geq T_m + D$, and N_R the smallest integer satisfying $N_R \Delta \geq T_m$. If $N \geq N_D$, $x(t_0) \in \Omega_\rho$ and $d_0 = 0$, then $x(t)$ is ultimately bounded in $\Omega_{\rho_d} \subseteq \Omega_\rho$ where:

$$\rho_d = \rho_{\min} + f_V\big(f_W(N_D \Delta)\big) + f_V\big(f_W(D)\big) \tag{4.101}$$

with $\rho_{\min}$ defined in Eq. 4.25.

Proof We assume that at t_a, a delayed measurement containing new information $x(t_a - d_a)$ is received, and that the next measurement with new state information is not received until t_{a+i}. This implies that $t_{a+i} - d_{a+i} > t_a - d_a$ and that the DMPC of Eqs. 4.81–4.90 is evaluated at t_a and the optimal input trajectories $u_{d1}^*(t|t_a)$ and $u_{d2}^*(t|t_a)$ are applied from t_a to t_{a+i} (see the input trajectories defined in Eqs. 4.98–4.99). We follow a similar approach as before; that is, to prove that $V(x(t_a))$ is a decreasing sequence of values with a lower bound.

Part 1: In this part, we prove that the stability results stated in Theorem 4.3 hold for $t_{a+i} - t_a = N_{D,a} \Delta$ and all $d_a \leq D$. By Proposition 2.1, the following inequality can be obtained:

$$V\big(\hat{x}(t_{a+i})\big) \leq \max\big\{V\big(\hat{x}(t_a)\big) - N_{D,a} \varepsilon_s, \rho_{\min}\big\}. \tag{4.102}$$

From the constraints of Eqs. 4.90 and 4.97 in LMPC 2 of Eq. 4.90 and LMPC 1 of Eq. 4.97, the following inequality can be written:

$$V\big(\check{x}(t)\big) \le V\big(\tilde{x}(t)\big) \le V\big(\hat{x}(t)\big), \quad \forall t \in [t_a, t_a + N_{D,a}\Delta). \tag{4.103}$$

From the inequalities of Eqs. 4.102, 4.103 and taking into account that $\hat{x}(t_a) = \check{x}(t_a) = \tilde{x}(t_a)$, the following inequality is obtained:

$$V\big(\check{x}(t_{a+i})\big) \le \max\big\{V\big(\tilde{x}(t_a)\big) - N_{D,a}\varepsilon_s, \rho_{\min}\big\}. \tag{4.104}$$

When $x(t) \in \Omega_\rho$ for all times (this point will be proved below), we can apply Proposition 2.3 to obtain the following inequalities:

$$V\big(\tilde{x}(t_a)\big) \le V\big(x(t_a)\big) + f_V\big(\|x(t_a) - \tilde{x}(t_a)\|\big), \tag{4.105}$$

$$V\big(x(t_{a+i})\big) \le V\big(\check{x}(t_{a+i})\big) + f_V\big(\|x(t_{a+i}) - \check{x}(t_{a+i})\|\big). \tag{4.106}$$

Applying Proposition 2.2, we obtain the following bounds on the deviation of $\tilde{x}(t)$ and $\check{x}(t)$ from $x(t)$:

$$\|x(t_a) - \tilde{x}(t_a)\| \le f_W(d_a), \tag{4.107}$$

$$\|x(t_{a+i}) - \check{x}(t_{a+i})\| \le f_W(N_D\Delta). \tag{4.108}$$

Note that Proposition 2.2 can be applied because the constraints of Eqs. 4.82, 4.83, 4.84, 4.87 and the implementation procedure guarantee that $\tilde{x}(t_a)$ and $\check{x}(t_{a+i})$ have been estimated using the same inputs applied to the system. We have also taken into account that $N_D\Delta \ge N_{D,a} + d_a$ for all d_a. Using the inequalities of Eqs. 4.104, 4.105–4.106 and 4.107–4.108, the following upper bound on $V(x(t_{a+i}))$ is obtained:

$$V\big(x(t_{a+i})\big) \le \max\big\{V\big(x(t_a)\big) - N_{D,a}\varepsilon_s, \rho_{\min}\big\} + f_V\big(f_W(N_D\Delta)\big) + f_V\big(f_W(d_a)\big). \tag{4.109}$$

In order to prove that the Lyapunov function is decreasing between two consecutive new measurements, the following inequality must hold:

$$N_{D,a}\varepsilon_s > f_V\big(f_W(N_D\Delta)\big) + f_V\big(f_W(d_a)\big) \tag{4.110}$$

for all possible $0 \le d_a \le D$. Taking into account that $f_W(\cdot)$ and $f_V(\cdot)$ are strictly increasing functions of their arguments, that $N_{D,a}$ is a decreasing function of the delay d_a and that if $d_a = D$ then $N_{D,a} = N_R$, if the condition of Eq. 4.100 is satisfied, the condition of Eq. 4.110 holds for all possible d_a and there exists $\varepsilon_w > 0$ such that the following inequality holds:

$$V\big(x(t_{a+i})\big) \le \max\big\{V\big(x(t_a)\big) - \varepsilon_w, \rho_d\big\}, \tag{4.111}$$

which implies that if $x(t_a) \in \Omega_\rho/\Omega_{\rho_d}$, then $V(x(t_{a+i})) < V(x(t_a))$, and if $x(t_a) \in \Omega_{\rho_d}$, then $V(x(t_{a+i})) \le \rho_d$.

Because the upper bound on the difference between the Lyapunov function of the actual trajectory x and the nominal trajectory $\check{x}$ is a strictly increasing function of time, the inequality of Eq. 4.111 also implies that:

$$V\big(x(t)\big) \leq \max\big\{V\big(x(t_a)\big), \rho_d\big\}, \quad \forall t \in [t_a, t_{a+i}). \tag{4.112}$$

Using the inequality of Eq. 4.112 recursively, it can be proved that if $x(t_0) \in \Omega_\rho$, then the closed-loop trajectories of the system of Eq. 4.1 under the DMPC of Eqs. 4.81–4.97 stay in Ω_ρ for all times (i.e., $x(t) \in \Omega_\rho$, $\forall t$). Moreover, using the inequality of Eq. 4.111 recursively, it can be proved that if $x(t_0) \in \Omega_\rho$, the closed-loop trajectories of the system of Eq. 4.1 under the DMPC of Eqs. 4.81–4.97 satisfy:

$$\limsup_{t \to \infty} V\big(x(t)\big) \leq \rho_d. \tag{4.113}$$

This proves that $x(t) \in \Omega_\rho$ for all times and that $x(t)$ is ultimately bounded in Ω_{ρ_d} when $t_{a+i} - t_a = N_{D,a}\Delta$ for all a.

Part 2: In this part, we extend the results proved in Part 1 to the general case, that is, $t_{a+i} - t_a \leq N_{D,a}\Delta$. Taking into account that $f_V(\cdot)$ and $f_W(\cdot)$ are strictly increasing functions of their arguments and following similar steps in Part 1, it is easy to prove that the inequality of Eq. 4.110 holds for all possible $d_a \leq D$ and $t_{a+i} - t_a \leq N_{D,a}\Delta$. Using this inequality and following the same line of argument as in the previous part, the stability results stated in Theorem 4.3 can be proved.   □

Remark 4.14 The sufficient conditions presented in Theorem 4.3 state that in order to guarantee practical stability, $V(x(t_a))$ must be a decreasing sequence of values with a lower bound for the worst possible case from a feedback control point of view; that is, the measurements are received every T_m (maximum time between successive measurements) with a delay equal to the maximum delay D.

Remark 4.15 In this section, we do not explicitly consider delays introduced in the system by the communication network or by the time needed to solve each of the LMPC optimization problems. Such delays are usually small (particularly in the context of DMPC) compared to the measurement delays and can be modeled as part of an overall measurement delay.

4.6.4 Application to a Reactor–Separator Process

We consider the reactor–separator process of Eqs. 3.52–3.63 described in Sect. 1.2.3 with the parameter values given in Table 3.3. In this subsection, we compare the performance of the DMPC of Eqs. 4.81–4.97 with that of the DMPC of Eqs. 4.49–4.62 in the case where the delayed state measurements of the process are available asynchronously at time instants $\{t_{a \geq 0}\}$. The same sampling time Δ and weighting matrices Q_c, R_c and R_{c2} used in Sect. 4.5.4 are used. The prediction horizons of

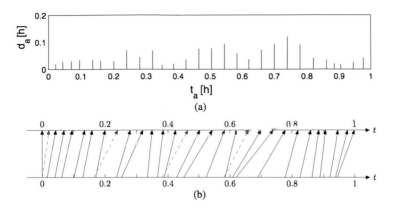

Fig. 4.15 Asynchronous time sequence $\{t_{a\geq0}\}$ and corresponding delay sequence $\{d_{a\geq0}\}$ with $T_m = 0.04$ h and $D = 0.12$ h: (**a**) the x-axis indicates $\{t_{a\geq0}\}$ and the y-axis indicates the size of d_a; (**b**) the *upper axis* indicates $\{t_{a\geq0}\}$, the *lower axis* indicates $t_a - d_a$, each *arrow* points from $t_a - d_a$ to corresponding t_a and the *dashed arrows* indicate the measurements which do not contain new information

both LMPC 1 and LMPC 2 are chosen to be $N = 8$ in this set of simulations so that the horizon covers the maximum possible open-loop operation interval. Note that the same estimated current state is used to evaluate both of the controllers.

The Poisson process used in Sect. 4.5.4 is used to generate $\{t_{a\geq0}\}$ with $W = 30$ and $T_m = 0.04$ h and another random process is used to generate the associated delay sequence $\{d_{a\geq0}\}$ with $D = 0.12$ h. Figure 4.15 shows the time instants when new state measurements are received, the associated delay sizes and the instants when the received measurements do not contain new information (which are discarded). The average time interval between two successive sampling times is 0.035 h and the average time delay is 0.057 h.

The state and input trajectories of the process of Eqs. 3.52–3.63 in closed-loop under the DMPC of Eqs. 4.81–4.97 and the DMPC of Eqs. 4.49–4.62 are shown in Figs. 4.16 and 4.17. In Fig. 4.16, we see that the DMPC of Eqs. 4.81–4.97 is able to stabilize the process at the desired steady state in about 0.6 h, but the control design of Eqs. 4.49–4.62 which does not account for measurement delays fails to drive the state to the desired steady state within 1 h.

Remark 4.16 We have also carried out simulations to evaluate the computational time of the LMPCs. The simulations have been carried out using MATLAB® in a PENTIUM® 3.20 GHz processor. The optimization problems have been solved using the built-in nonlinear programming function *fmincom* of MATLAB®. For 50 evaluations, the mean time to solve LMPC 2 of Eqs. 4.49–4.55 and LMPC 1 of Eqs. 4.56–4.62 are 5.52 seconds and 2.90 seconds, respectively, with the prediction horizon $N = 6$; the mean time to solve LMPC 2 of Eqs. 4.81–4.90 and LMPC 1 of Eqs. 4.91–4.97 are 13.95 seconds and 6.83 seconds, respectively, with the prediction horizon $N = 8$. These computational times can be reduced significantly by using a

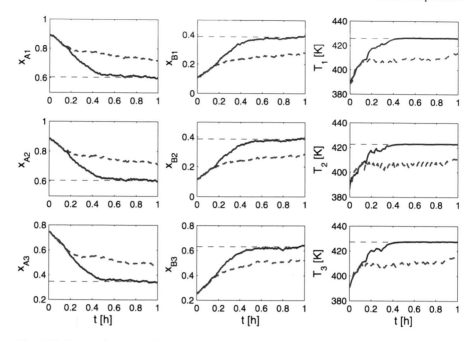

Fig. 4.16 State trajectories of the reactor–separator process of Eqs. 3.52–3.63 under the DMPC of Eqs. 4.81–4.97 (*solid lines*) and the DMPC of Eqs. 4.49–4.62 (*dashed lines*) in the presence of asynchronous and delayed measurements

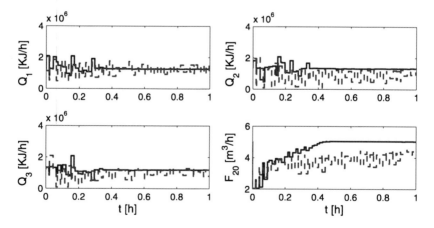

Fig. 4.17 Input trajectories of the reactor–separator process of Eqs. 3.52–3.63 under the DMPC of Eqs. 4.81–4.97 (*solid lines*) and the DMPC of Eqs. 4.49–4.62 (*dashed lines*) in the presence of asynchronous and delayed measurements

compiled nonlinear programming solver implemented in *C* or other programming languages.

4.7 Conclusions

In this chapter, we focused on a class of nonlinear control problems that arise when new control systems which may use networked sensors and/or actuators are added to already operating control loops to improve closed-loop performance. To address this control problem, a DMPC method was introduced where both the preexisting control system and the new control system are designed via LMPC theory. The presented DMPC design stabilizes the closed-loop system, improves the closed-loop performance and allows handling input constraints. In addition, the distributed control design requires reduced communication between the two distributed controllers since it requires that these controllers communicate only once at each sampling time and is computationally more efficient compared to the corresponding centralized model predictive control design. In addition, the DMPC method is also extended to include nonlinear systems subject to asynchronous and delayed measurements. Extensive simulations using a chemical plant network example, described by a nonlinear model, demonstrated the applicability and effectiveness of the DMPC designs.

Chapter 5
Distributed Model Predictive Control: Multiple-Controller Cooperation

5.1 Introduction

In Chap. 4, we presented a DMPC architecture with one-directional communication for a very broad class of nonlinear systems. In this architecture, two separate controllers designed via LMPC were considered, in which one LMPC was used to guarantee the stability of the closed-loop system and the other LMPC was used to improve the closed-loop performance. In this chapter, we focus on DMPC of large-scale nonlinear systems in which several distinct sets of manipulated inputs are used to regulate the system. For each set of manipulated inputs, a different model predictive controller, which is able to communicate with the rest of the controllers in making its decisions, is used to compute the control actions. Specifically, under the assumption that feedback of the state of the process is available to all the distributed controllers at each sampling time and that a model of the plant is available, we present two different DMPC architectures designed via LMPC techniques. In the first architecture, the distributed controllers use a one-directional communication strategy, are evaluated in sequence and each controller is evaluated only once at each sampling time; in the second architecture, the distributed controllers utilize a bidirectional communication strategy, are evaluated in parallel and iterate to improve closed-loop performance. In order to ensure the stability of the closed-loop system, each model predictive controller in both architectures incorporates a stability constraint which is based on a suitable nonlinear control law which can stabilize the closed-loop system. We prove that the two DMPC architectures enforce practical stability in the closed-loop system while improving performance.

Moreover, the DMPC designs will be also extended to include nonlinear systems subject to asynchronous and delayed state feedback. In the case of asynchronous feedback, under the assumption that there is an upper bound on the maximum interval between two consecutive measurements, we first extend both the DMPC architectures to take explicitly into account asynchronous feedback. Subsequently, we design a DMPC scheme using bi-directional communication for systems subject to asynchronous measurements that also involve time-delays under the assumption that there exists an upper bound on the maximum feedback delay. Sufficient conditions under which the proposed distributed control designs guarantee that the states

P.D. Christofides et al., *Networked and Distributed Predictive Control*, Advances in Industrial Control, DOI 10.1007/978-0-85729-582-8_5, © Springer-Verlag London Limited 2011

of the closed-loop system are ultimately bounded in regions that contain the origin are provided. The theoretical results are illustrated through a catalytic alkylation of benzene process example.

Finally, in this chapter, we will focus on a hierarchical type DMPC and discuss how to handle communication disruptions—communication channel noise and data losses—between the distributed controllers. To handle communication disruptions, feasibility problems are incorporated in the DMPC architecture to determine whether the data transmitted through the communication channel is reliable or not. Based on the results of the feasibility problems, the transmitted information is accepted or rejected by the stabilizing MPC. In order to ensure the stability of the closed-loop system under communication disruptions, each model predictive controller utilizes a suitable Lyapunov-based stability constraint. The results of this chapter were first presented in [9, 30, 54, 55, 58].

5.2 System Description

In this chapter, we consider nonlinear systems described by the following state-space model:

$$\dot{x}(t) = f\big(x(t)\big) + \sum_{i=1}^{m} g_i\big(x(t)\big)u_i(t) + k\big(x(t)\big)w(t), \qquad (5.1)$$

where $x(t) \in R^n$ denotes the vector of state variables, $u_i(t) \in R^{m_i}, i = 1, \ldots, m$, are m sets of control (manipulated) inputs and $w(t) \in R^w$ denotes the vector of disturbance variables. The m sets of inputs are restricted to be in m nonempty convex sets $U_i \subseteq R^{m_{u_i}}, i = 1, \ldots, m$, which are defined as follows:

$$U_i := \big\{u_i \in R^{m_i} : \|u_i\| \leq u_i^{\max}\big\}, \quad i = 1, \ldots, m, \qquad (5.2)$$

where $u_i^{\max}, i = 1, \ldots, m$, are the magnitudes of the input constraints. The disturbance vector is bounded, i.e., $w(t) \in W$ where:

$$W := \big\{w \in R^w : \|w\| \leq \theta, \theta > 0\big\} \qquad (5.3)$$

with θ being a known positive real number.

We assume that $f, g_i, i = 1, \ldots, m$, and k are locally Lipschitz vector, matrix and matrix functions, respectively, and that the origin is an equilibrium of the unforced nominal system (i.e., the system of Eq. 5.1 with $u_i(t) = 0, i = 1, \ldots, m, w(t) = 0$ for all t) which implies that $f(0) = 0$.

Remark 5.1 In this chapter, in order to account for DMPC designs in which the distributed controllers are evaluated in parallel, we consider nonlinear systems with control inputs entering the system dynamics in an affine fashion. We note that the results presented in Sects. 5.4.1 and 5.5.2 can be extended to more general nonlinear systems, for example, systems described by the following state-space model:

$$\dot{x}(t) = f\big(x(t), u_1(t), \ldots, u_m(t), w(t)\big). \tag{5.4}$$

5.3 Lyapunov-Based Control

We assume that there exists a nonlinear control law $h(x) = [h_1(x)^T \cdots h_m(x)^T]^T$ with $u_i = h_i(x)$, $i = 1, \ldots, m$, which renders (under continuous state feedback) the origin of the nominal closed-loop system asymptotically stable while satisfying the input constraints for all the states x inside a given stability region. Using converse Lyapunov theorems, this assumption implies that there exist functions $\alpha_i(\cdot)$, $i = 1, 2, 3, 4$ of class $\mathcal{K}$ and a continuously differentiable Lyapunov function $V(x)$ for the nominal closed-loop system that satisfy the following inequalities:

$$\alpha_1\big(\|x\|\big) \leq V(x) \leq \alpha_2\big(\|x\|\big), \tag{5.5}$$

$$\frac{\partial V(x)}{\partial x}\left(f(x) + \sum_{i=1}^{m} g_i(x)h_i(x)\right) \leq -\alpha_3\big(\|x\|\big), \tag{5.6}$$

$$\left\|\frac{\partial V(x)}{\partial x}\right\| \leq \alpha_4\big(\|x\|\big), \tag{5.7}$$

$$h_i(x) \in U_i, \quad i = 1, \ldots, m \tag{5.8}$$

for all $x \in O \subseteq R^{n_x}$ where O is an open neighborhood of the origin. We denote the region $\Omega_\rho \subseteq O$ as the stability region of the closed-loop system under the nonlinear control law $h(x)$.

By continuity, the local Lipschitz property assumed for the vector fields $f(x)$, $g_i(x)$, $i = 1, \ldots, m$, and $k(x)$ and taking into account that the manipulated inputs u_i, $i = 1, \ldots, m$, and the disturbance w are bounded in convex sets, there exist positive constants M, M_{g_i}, L_x, L_{u_i} and L_w $(i = 1, \ldots, m)$ such that:

$$\left\|f(x) + \sum_{i=1}^{m} g_i(x)u_i + k(x)w\right\| \leq M, \tag{5.9}$$

$$\|g_i(x)\| \leq M_{g_i}, \quad i = 1, \ldots, m, \tag{5.10}$$

$$\|f(x) - f(x')\| \leq L_x \|x - x'\|, \tag{5.11}$$

$$\|g_i(x) - g_i(x')\| \leq L_{u_i} \|x - x'\|, \quad i = 1, \ldots, m, \tag{5.12}$$

$$\|k(x)\| \leq L_w \tag{5.13}$$

for all $x, x' \in \Omega_\rho$, $u_i \in U_i$, $i = 1, \ldots, m$, and $w \in W$. In addition, by the continuous differentiable property of the Lyapunov function $V(x)$, there exist positive constants

L'_x, L'_{u_i}, $i = 1, \ldots, m$, and L'_w such that:

$$\left\| \frac{\partial V(x)}{\partial x} f(x) - \frac{\partial V(x')}{\partial x} f(x') \right\| \leq L'_x \|x - x'\|, \tag{5.14}$$

$$\left\| \frac{\partial V(x)}{\partial x} g_i(x) - \frac{\partial V(x')}{\partial x} g_i(x') \right\| \leq L'_{u_i} \|x - x'\|, \quad i = 1, \ldots, m, \tag{5.15}$$

$$\left\| \frac{\partial V(x)}{\partial x} k(x) \right\| \leq L'_w \tag{5.16}$$

for all $x, x' \in \Omega_\rho$, $u_i \in U_i$, $i = 1, \ldots, m$, and $w \in W$.

5.4 Sequential and Iterative DMPC Designs with Synchronous Measurements

The objective of this section is to design DMPC architectures including multiple MPCs for large-scale nonlinear process systems with continuous, synchronous state feedback. Specifically, we will discuss two different DMPC architectures. The first DMPC architecture is a direct extension of the DMPC presented in Sect. 4.4 in which different MPC controllers are evaluated in sequence, only once at each sampling time and require only one-directional communication between consecutive distributed controllers (i.e., the distributed controllers are connected by pairs). In the second architecture, different MPCs are evaluated in parallel, once or more than once at each sampling time depending on the number of iterations, and bidirectional communication among all the distributed controllers (i.e., the distributed controllers are all interconnected) is used.

In each DMPC architecture, we will design m LMPCs to compute u_i, $i = 1, \ldots, m$, and refer to the LMPC computing the input trajectories of u_i as LMPC i. In addition, we assume that the state x of the system of Eq. 5.1 is sampled synchronously and the time instants at which we have state measurement samplings are indicated by the time sequence $\{t_{k \geq 0}\}$ with $t_k = t_0 + k\Delta$, $k = 0, 1, \ldots$ where t_0 is the initial time and Δ is the sampling time. The results will be extended to include systems subject to asynchronous and delayed measurements in Sects. 5.5 and 5.6.

5.4.1 Sequential DMPC

A schematic of the architecture considered in this subsection is shown in Fig. 5.1.

5.4.1.1 Sequential DMPC Formulation

We first present the implementation strategy of this DMPC architecture and then design the corresponding LMPCs. The implementation strategy of this DMPC architecture is as follows:

Fig. 5.1 Sequential DMPC architecture

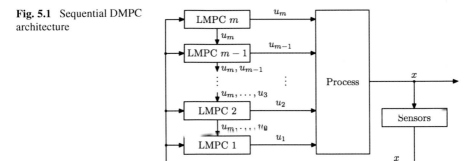

1. At t_k, all the LMPCs receive the state measurement $x(t_k)$ from the sensors.
2. For $j = m$ to 1
    2.1. LMPC j receives the entire future input trajectories of u_i, $i = m, \ldots, j+1$, from LMPC $j+1$ and evaluates the future input trajectory of u_j based on $x(t_k)$ and the received future input trajectories.
    2.2. LMPC j sends the first step input value of u_j to its actuators and the entire future input trajectories of u_i, $i = m, \ldots, j$, to LMPC $j-1$.
3. When a new measurement is received ($k \leftarrow k+1$), go to Step 1.

In this architecture, each LMPC only sends its future input trajectory and the future input trajectories it received to the next LMPC (i.e., LMPC j sends input trajectories to LMPC $j-1$). This implies that LMPC j, $j = m, \ldots, 2$, does not have any information about the values that u_i, $i = j-1, \ldots, 1$ will take when the optimization problems of the LMPCs are designed. In order to make a decision, LMPC j, $j = m, \ldots, 2$ must assume trajectories for u_i, $i = j-1, \ldots, 1$, along the prediction horizon. To this end, the nonlinear control law $h(x)$ is used. In order to inherit the stability properties of the controller $h(x)$, each control input u_i, $i = 1, \ldots, m$ must satisfy a constraint that guarantees a given minimum contribution to the decrease rate of the Lyapunov function $V(x)$. Specifically, the design of LMPC j, $j = 1, \ldots, m$, is based on the following optimization problem:

$$\min_{u_j \in S(\Delta)} \int_{t_k}^{t_k+N} \left[\left\| \tilde{x}(\tau) \right\|_{Q_c} + \sum_{i=1}^{m} \left\| u_i(\tau) \right\|_{R_{ci}} \right] d\tau, \tag{5.17}$$

$$\text{s.t.} \quad \dot{\tilde{x}}(t) = f\big(\tilde{x}(t)\big) + \sum_{i=1}^{m} g_i\big(\tilde{x}(t)\big) u_i, \tag{5.18}$$

$$u_i(t) = h_i\big(\tilde{x}(t_{k+l})\big),$$
$$i = 1, \ldots, j-1, \forall t \in [t_{k+l}, t_{k+l+1}), l = 0, \ldots, N-1, \tag{5.19}$$

$$u_i(t) = u_{s,i}^*(t|t_k), \quad i = j+1, \ldots, m, \tag{5.20}$$

$$u_j(t) \in U_j, \tag{5.21}$$

$$\tilde{x}(t_k) = x(t_k), \tag{5.22}$$

$$\frac{\partial V(x(t_k))}{\partial x}g_j\big(x(t_k)\big)u_j(t_k) \leq \frac{\partial V(x(t_k))}{\partial x}g_j\big(x(t_k)\big)h_j\big(x(t_k)\big). \qquad (5.23)$$

In the optimization problem of Eqs. 5.17–5.23, $u_{s,i}^*(t|t_k)$ denotes the optimal future input trajectory of u_i obtained by LMPC i of the form of Eqs. 5.17–5.23 evaluated before LMPC j, $\tilde{x}$ is the predicted trajectory of the nominal system with $u_i = u_{s,i}$, $i = j+1, \ldots, m$, u_i, $i = 1, \ldots, j-1$, the corresponding elements of $h(x)$ applied in a sample-and-hold fashion and u_j predicted by LMPC j of Eqs. 5.17–5.23. The optimal solution to the optimization problem of Eqs. 5.17–5.23 is denoted as $u_{s,j}^*(t|t_k)$ which is defined for $t \in [t_k, t_{k+N})$.

The constraint of Eq. 5.18 is the nominal model of the system of Eq. 5.1, which is used to predict the future evolution of the system; the constraint of Eq. 5.19 defines the value of the inputs evaluated after u_j (i.e., u_i with $i = 1, \ldots, j-1$); the constraint of Eq. 5.20 defines the value of the inputs evaluated before u_j (i.e., u_i with $i = j+1, \ldots, m$); the constraint of Eq. 5.21 is the constraint on the manipulated input u_j; the constraint of Eq. 5.22 sets the initial state for the optimization problem; the constraint of Eq. 5.23 guarantees that the contribution of input u_j to the decrease rate of the time derivative of the Lyapunov function $V(x)$ at the initial evaluation time (i.e., at t_k), if $u_j = u_{s,j}^*(t_k|t_k)$ is applied, is bigger than or equal to the value obtained when $u_j = h_j(x(t_k))$ is applied. This constraint allows proving the closed-loop stability properties of this DMPC.

The manipulated inputs of the system of Eq. 5.1 under the DMPC are defined as follows:

$$u_i(t) = u_{s,i}^*(t|t_k), \quad i = 1, \ldots, m, \forall t \in [t_k, t_{k+1}). \qquad (5.24)$$

In what follows, we refer to this DMPC architecture as the sequential DMPC.

Remark 5.2 Note that, in order to simplify the description of the implementation strategy presented above in this subsection, we do not distinguish LMPC m and LMPC 1 from the others. We note that LMPC m does not receive any information from the other controllers and LMPC 1 does not have to send information to any other controller.

Remark 5.3 Note also that the assumption that the full state x of the system is sampled synchronously is a widely used assumption in the control system design. The control system designs presented in this section can be extended to the case where only part of the state x is measurable by designing an observer to estimate the whole state vector from output measurements and by designing the control system based on the measured and estimated states. In this case, the stability properties of the resulting output feedback control systems are affected by the convergence of the observer and need to be carefully studied.

5.4.1.2 Stability Properties

The sequential DMPC of Eqs. 5.17–5.24 computes the inputs u_i, $i = 1, \ldots, m$, applied to the system of Eq. 5.1 in a way such that in the closed-loop system, the value of the Lyapunov function at time instant t_k (i.e., $V(x(t_k))$) is a decreasing sequence of values with a lower bound. Following Lyapunov arguments, this property guarantees practical stability of the closed-loop system. This is achieved due to the constraint of Eq. 5.23. This property is presented in Theorem 5.1 below.

Theorem 5.1 *Consider the system of Eq. 5.1 in closed-loop under the sequential DMPC of Eqs. 5.17–5.24 based on a nonlinear control law $h(x)$ that satisfies the condition of Eqs. 5.5–5.8 with class $\mathcal{K}$ functions $\alpha_i(\cdot)$, $i = 1, 2, 3, 4$. Let $\varepsilon_w > 0$, $\Delta > 0$ and $\rho > \rho_s > 0$ satisfy the following constraint:*

$$-\alpha_3\left(\alpha_2^{-1}(\rho_s)\right) + L^* \leq -\varepsilon_w/\Delta, \tag{5.25}$$

where $L^* = (L'_x + \sum_{i=1}^{m} L'_{u_i} u_i^{\max})M + L'_w\theta$ with M, L'_x, L'_{u_i} $(i = 1, \ldots, m)$ and L'_w defined in Eqs. 5.9–5.16. For any $N \geq 1$, if $x(t_0) \in \Omega_\rho$ and if $\rho_{\min} \leq \rho$ where:

$$\rho_{\min} = \max\left\{V\left(x(t + \Delta)\right) : V\left(x(t)\right) \leq \rho_s\right\}, \tag{5.26}$$

then the state $x(t)$ of the closed-loop system is ultimately bounded in $\Omega_{\rho_{\min}}$.

Proof The proof consists of two parts. We first prove that the optimization problem of Eqs. 5.17–5.23 is feasible for all $j = 1, \ldots, m$ and $x \in \Omega_\rho$. Then we prove that, under the DMPC of Eqs. 5.17–5.24, the state of the system of Eq. 5.1 is ultimately bounded in $\Omega_{\rho_{\min}}$. Note that the constraint of Eq. 5.23 of each distributed controller is independent from the decisions that the rest of the distributed controllers make.

Part 1: In order to prove the feasibility of the optimization problem of Eqs. 5.17–5.23, we only have to prove that there exists a $u_j(t_k)$ which satisfies the input constraint of Eq. 5.21 and the constraint of Eq. 5.23. This is because the constraint of Eq. 5.23 is only enforced on the first prediction step of $u_j(t)$ and does not depend on the values of the inputs chosen by the rest of the controllers (see Remark 5.9). In the prediction time $t \in [t_{k+1}, t_{k+N})$, the input constraint of Eq. 5.24 can be easily satisfied with $u_j(\tau)$ being any value in the convex set U_j.

We assume that $x(t_k) \in \Omega_\rho$ ($x(t)$ is bounded in Ω_ρ which will be proved in Part 2). It is easy to verify that the value of u_j such that $u_j(t_k) = h_j(x(t_k))$ satisfies the input constraint of Eq. 5.21 (assumed property of $h(x)$ for $x \in \Omega_\rho$) and the constraint of Eq. 5.23, thus, the feasibility of the optimization problem of LMPC j of Eqs. 5.17–5.23, $j = 1, \ldots, m$, is guaranteed.

Part 2: From the condition of Eq. 5.6 and the constraint of Eq. 5.23, if $x(t_k) \in \Omega_\rho$, it follows that:

$$\frac{\partial V(x(t_k))}{\partial x}\left(f\left(x(t_k)\right) + \sum_{i=1}^{m} g_i\left(x(t_k)\right)u_{s,i}^*(t_k|t_k)\right)$$

$$\leq \frac{\partial V(x(t_k))}{\partial x}\left(f\left(x(t_k)\right) + \sum_{i=1}^{m} g_i\left(x(t_k)\right)h_i\left(x(t_k)\right)\right)$$

$$\leq -\alpha_3\left(\|x(t_k)\|\right). \tag{5.27}$$

The time derivative of the Lyapunov function V along the actual state trajectory $x(t)$ of the system of Eq. 5.1 in $t \in [t_k, t_{k+1})$ is given by:

$$\dot{V}\big(x(t)\big) = \frac{\partial V(x(t))}{\partial x}\left(f\big(x(t)\big) + \sum_{i=1}^{m} g_i\big(x(t)\big)u_{s,i}^*(t_k|t_k) + k\big(x(t)\big)w(t)\right). \quad (5.28)$$

Adding and subtracting $\frac{\partial V(x(t_k))}{\partial x}(f(x(t_k)) + \sum_{i=1}^{m} g_i(x(t_k))u_{s,i}^*(t_k|t_k))$ and taking into account Eq. 5.27, we obtain the following inequality:

$$\dot{V}\big(x(t)\big) \leq -\alpha_3\big(\|x(t_k)\|\big)$$
$$+ \frac{\partial V(x(t))}{\partial x}\left(f\big(x(t)\big) + \sum_{i=1}^{m} g_i\big(x(t)\big)u_{s,i}^*(t_k|t_k) + k\big(x(t)\big)w(t)\right)$$
$$- \frac{\partial V(x(t_k))}{\partial x}\left(f\big(x(t_k)\big) + \sum_{i=1}^{m} g_i\big(x(t_k)\big)u_{s,i}^*(t_k|t_k)\right). \quad (5.29)$$

Taking into account Eqs. 5.5 and 5.9, the following inequality if obtained for all $x(t_k) \in \Omega_\rho / \Omega_{\rho_s}$ from Eq. 5.29:

$$\dot{V}\big(x(t)\big) \leq -\alpha_3\big(\alpha_2^{-1}(\rho_s)\big) + \left(L_x' + \sum_{i=1}^{m} L_{u_i}' u_{s,i}^*(t_k|t_k)\right)\|x(t) - x(t_k)\| + L_w'\|w(t)\|.$$
$$(5.30)$$

Taking into account Eq. 5.9 and the continuity of $x(t)$, the following bound can be written for all $t \in [t_k, t_{k+1})$:

$$\|x(t) - x(t_k)\| \leq M\Delta. \quad (5.31)$$

Using this expression, the bounds on the disturbance $w(t)$ and the inputs u_i, $i = 1, \ldots, m$, and Eq. 5.30, we obtain the following bound on the time derivative of the Lyapunov function for $t \in [t_k, t_{k+1})$, for all initial states $x(t_k) \in \Omega_\rho / \Omega_{\rho_s}$:

$$\dot{V}\big(x(t)\big) \leq -\alpha_3\big(\alpha_2^{-1}(\rho_s)\big) + \left(L_x' + \sum_{i=1}^{m} L_{u_i}' u_i^{\max}\right) M + L_w'\theta. \quad (5.32)$$

If the condition of Eq. 5.25 is satisfied, then there exists $\varepsilon_w > 0$ such that the following inequality holds for $x(t_k) \in \Omega_\rho / \Omega_{\rho_s}$:

$$\dot{V}\big(x(t)\big) \leq -\varepsilon_w / \Delta \quad (5.33)$$

for $t \in [t_k, t_{k+1})$. Integrating the inequality of Eq. 5.33 on $t \in [t_k, t_{k+1})$, we obtain that:

$$V\big(x(t_{k+1})\big) \leq V\big(x(t_k)\big) - \varepsilon_w, \quad (5.34)$$
$$V\big(x(t)\big) \leq V\big(x(t_k)\big), \quad \forall t \in [t_k, t_{k+1}) \quad (5.35)$$

Fig. 5.2 Iterative DMPC
architecture

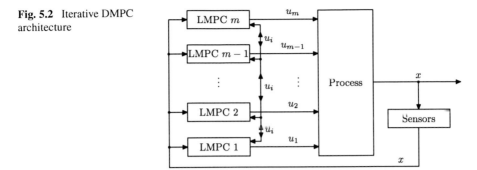

for all $x(t_k) \in \Omega_\rho/\Omega_{\rho_s}$. Using Eqs. 5.34 and 5.35 recursively it can be proved that, if $x(t_0) \in \Omega_\rho/\Omega_{\rho_s}$, the state converges to Ω_{ρ_s} in a finite number of sampling times without leaving the stability region. Once the state converges to $\Omega_{\rho_s} \subseteq \Omega_{\rho_{\min}}$, it remains inside $\Omega_{\rho_{\min}}$ for all times. This statement holds because of the definition of $\rho_{\min}$. This proves that the closed-loop system under the sequential DMPC of Eqs. 5.17–5.24 is ultimately bounded in $\Omega_{\rho_{\min}}$. □

Remark 5.4 The sequential DMPC approach can be applied to more general non-linear systems as described in Eq. 5.4 (see Remark 5.1) by a proper redesign of the Lyapunov-based constraints of Eqs. 5.23 ($j = 1, \ldots, m$) following the method used in the design of the constraints of Eq. 4.15 and 4.21, see Sect. 4.4.1.

5.4.2 Iterative DMPC

An alternative architecture to the sequential DMPC architecture presented in the previous subsection is to evaluate all the distributed LMPCs in parallel and iterate to improve closed-loop performance. A schematic of this control architecture is shown in Fig. 5.2.

5.4.2.1 Iterative DMPC Formulation

In this architecture, each distributed LMPC must be able to communicate with all the other controllers (i.e., the distributed controllers are all interconnected). More specifically, when a new state measurement is available at a sampling time, each distributed LMPC controller evaluates and obtains its future input trajectory; and then each LMPC controller broadcasts its latest obtained future input trajectory to all the other controllers. Based on the newly received input trajectories, each LMPC controller evaluates its future input trajectory again and this process is repeated until a certain termination condition is satisfied. Specifically, the implementation strategy is as follows:

1. At t_k, all the LMPCs receive the state measurement $x(t_k)$ from the sensors and then evaluate their future input trajectories in an iterative fashion with initial input guesses generated by $h(\cdot)$.
2. At iteration c $(c \geq 1)$:
    2.1. Each LMPC evaluates its own future input trajectory based on $x(t_k)$ and the latest received input trajectories of all the other LMPCs (when $c = 1$, initial input guesses generated by $h(\cdot)$ are used).
    2.2. The controllers exchange their future input trajectories. Based on all the input trajectories, each controller calculates and stores the value of the cost function.
3. If a termination condition is satisfied, each controller sends its entire future input trajectory corresponding to the smallest value of the cost function to its actuators; if the termination condition is not satisfied, go to Step 2 ($c \leftarrow c + 1$).
4. When a new measurement is received, go to Step 1 ($k \leftarrow k + 1$).

Note that at the initial iteration, all the LMPCs use $h(x)$ to estimate the input trajectories of all the other controllers. Note also that the number of iterations c can be variable and it does not affect the closed-loop stability of the DMPC architecture presented in this subsection; a point that will be made clear below. For the iterations in this DMPC architecture, there are different choices of the termination condition. For example, the number of iterations c may be restricted to be smaller than a maximum iteration number $c_{\max}$ (i.e., $c \leq c_{\max}$) and/or the iterations may be terminated when the difference of the performance or the solution between two consecutive iterations is smaller than a threshold value and/or the iterations maybe terminated when a maximum computational time is reached.

In order to proceed, we define $\hat{x}(t|t_k)$ for $t \in [t_k, t_{k+N})$ as the nominal sampled trajectory of the system of Eq. 5.1 associated with the feedback control law $h(x)$ and sampling time Δ starting from $x(t_k)$. This nominal sampled trajectory is obtained by integrating recursively the following differential equation:

$$\dot{\hat{x}}(t|t_k) = f\big(\hat{x}(t|t_k)\big) + \sum_{i=1}^{m} g_i\big(\hat{x}(t|t_k)\big) h_i\big(\hat{x}(t_{k+l}|t_k)\big),$$

$$\forall \tau \in [t_{k+l}, t_{k+l+1}), l = 0, \ldots, N - 1. \tag{5.36}$$

Based on $\hat{x}(t|t_k)$, we can define the following variable:

$$u_{n,j}(t|t_k) = h_j\big(\hat{x}(t_{k+l}|t_k)\big),$$

$$j = 1, \ldots, m, \forall \tau \in [t_{k+l}, t_{k+l+1}), l = 0, \ldots, N - 1, \tag{5.37}$$

which will be used as the initial guess of the trajectory of u_j.

The design of the LMPC j, $j = 1, \ldots, m$, at iteration c is based on the following optimization problem:

$$\min_{u_j \in S(\Delta)} \int_{t_k}^{t_{k+N}} \left[\big\| \tilde{x}(\tau) \big\|_{Q_c} + \sum_{i=1}^{m} \big\| u_i(\tau) \big\|_{R_{ci}} \right] d\tau, \tag{5.38}$$

$$\text{s.t.} \quad \dot{\tilde{x}}(t) = f\big(\tilde{x}(t)\big) + \sum_{i=1}^{m} g_i\big(\tilde{x}(t)\big)u_i, \tag{5.39}$$

$$u_i(t) = u_{p,i}^{*,c-1}(t|t_k), \quad \forall i \neq j, \tag{5.40}$$

$$u_j(t) \in U_j, \tag{5.41}$$

$$\tilde{x}(t_k) = x(t_k), \tag{5.42}$$

$$\frac{\partial V(x(t_k))}{\partial x} g_j\big(x(t_k)\big)u_j(t_k) \leq \frac{\partial V(x(t_k))}{\partial x} g_j\big(x(t_k)\big)h_j\big(x(t_k)\big), \tag{5.43}$$

where $\tilde{x}$ is the predicted trajectory of the nominal system with u_k, the input trajectory, computed by the LMPCs of Eqs. 5.38–5.43 and all the other inputs are the optimal input trajectories at iteration $c - 1$ of the rest of distributed controllers (i.e., $u_{p,i}^{*,c-1}(t|t_k)$ for $i \neq j$). The optimal solution to the optimization problem of Eqs. 5.38–5.43 is denoted as $u_{p,j}^{*,c}(t|t_k)$ which is defined for $t \in [t_k, t_{k+N})$. Accordingly, we define the final optimal input trajectory of LMPC j (that is, the optimal trajectories computed at the last iteration) as $u_{p,j}^{*}(t|t_k)$ which is also defined for $t \in [t_k, t_{k+N})$.

Note that in the first iteration of each distributed LMPC, the input trajectory defined in Eq. 5.37 is used as the initial input trajectory guess; that is, $u_{p,j}^{*,0}(t|t_k) = u_{n,j}(t|t_k)$ with $i = 1, \ldots, m$.

The manipulated inputs of the system of Eq. 5.1 under this DMPC design with LMPCs of Eqs. 5.38–5.43 are defined as follows:

$$u_i(t) = u_{p,i}^{*}(t|t_k), \quad i = 1, \ldots, m, \forall t \in [t_k, t_{k+1}). \tag{5.44}$$

In what follows, we refer to this DMPC architecture as the iterative DMPC. The stability properties of the iterative DMPC are stated in the following Theorem 5.2.

Remark 5.5 In general, there is no guaranteed convergence of the optimal cost or solution of an iterated DMPC to the optimal cost or solution of a centralized MPC for general nonlinear constrained systems because of the nonconvexity of the MPC optimization problems and the fact that the DMPC does not solve the centralized LMPC in a distributed fashion due to the way the Lyapunov-based constraint of the centralized LMPC is broken down into constraints imposed on the individual LMPCs; please also see Remark 5.12 below. However, with the implementation strategy of the iterative DMPC presented in this section, it is guaranteed that the optimal cost of the distributed optimization of Eqs. 5.38–5.43 is upper bounded by the cost of the Lyapunov-based controller $h(\cdot)$ at each sampling time.

Remark 5.6 Note that in the case of linear systems, the constraint of Eq. 5.54 is linear with respect to u_j and it can be verified that the optimization problem of Eqs. 5.50–5.54 is convex. The input given by LMPC j of Eqs. 5.50–5.54 at each iteration may be defined as a convex combination of the current optimal input solution and the previous one, for example,

$$u^c_{p,j}(t|t_k) = \sum_{i=1}^{m,i \neq j} w_i u^{c-1}_{p,j}(t|t_k) + w_j u^{*,c}_{p,j}(t|t_k), \qquad (5.45)$$

where $\sum_{i=1}^{m} w_i = 1$ with $0 < w_i < 1$, $u^{*,c}_{p,j}$ is the current solution given by the optimization problem of Eqs. 5.50–5.54 and $u^{c-1}_{p,j}$ is the convex combination of the solutions obtained at iteration $c - 1$. By doing this, it is possible to proved that the optimal cost of the distributed LMPC of Eqs. 5.50–5.54 converges to the one of the corresponding centralized control system [5, 98]. This property is summarized in Corollary 5.1 in Sect. 5.4.2.2. We also note that in the case of linear systems, the convexity of the distributed optimization problem also holds for all the other DMPC designs presented in this chapter and Chap. 6. In addition to Corollary 5.1, the reader may also refer to [5, 8, 93, 98] for more discussions on the conditions under which convergence of the solution of a distributed linear or convex MPC design to the solution of a centralized MPC or a Pareto optimal solution is ensured in the context of linear systems.

5.4.2.2 Stability Properties

Theorem 5.2 *Consider the system of Eq. 5.1 in closed-loop under the sequential DMPC of Eqs. 5.38–5.44 based on a nonlinear control law $h(x)$ that satisfies the condition of Eqs. 5.5–5.8 with class $\mathcal{K}$ functions $\alpha_i(\cdot)$, $i = 1, 2, 3, 4$. Let $\varepsilon_w > 0$, $\Delta > 0$ and $\rho > \rho_s > 0$ satisfy the constraint of Eq. 5.25. For any $N \geq 1$ and $c \geq 1$, if $x(t_0) \in \Omega_\rho$ and if $\rho_{\min} \leq \rho$ where $\rho_{\min}$ is defined as in Eq. 5.26, then the state $x(t)$ of the closed-loop system is ultimately bounded in $\Omega_{\rho_{\min}}$.*

Proof Similar to the proof of Theorem 5.1, the proof of Theorem 5.2 also consists of two parts. We first prove that the optimization problem of Eqs. 5.38–5.43 is feasible for each iteration c and $x \in \Omega_\rho$. Then we prove that, under the DMPC architecture of Eqs. 5.38–5.44, the state of the system of Eq. 5.1 is ultimately bounded in $\Omega_{\rho_{\min}}$.

Part 1: In order to prove the feasibility of the optimization problem of Eqs. 5.38–5.43, we only have to prove that there exists a $u_j(t_k)$ which satisfies the input constraint of Eq. 5.41 and the constraint of Eq. 5.43. This is because the constraint of Eq. 5.43 is only enforced on the first prediction step of $u_j(t_k)$ and in the prediction time $t \in [t_{k+1}, t_{k+N})$, the input constraint of Eq. 5.44 can be easily satisfied with $u_j(t)$ being any value in the convex set U_j.

We assume that $x(t_k) \in \Omega_\rho$ ($x(t)$ is bounded in Ω_ρ which will be proved in Part 2). It is easy to verify that the value of u_j such that $u_j(t_k) = h_j(x(t_k))$ satisfies the input constraint of Eq. 5.41 (assumed property of $h(x)$ for $x \in \Omega_\rho$) and the constraint of Eq. 5.43 for all possible c, thus, the feasibility of LMPC j of Eqs. 5.38–5.43, $j = 1, \ldots, m$, is guaranteed.

Part 2: By adding the constraint of Eq. 5.43 of each LMPC together, we have:

$$\sum_{j=1}^{m} \frac{\partial V(x(t_k))}{\partial x} g_j\big(x(t_k)\big) u_{p,j}^{*,c}(t_k|t_k) \leq \sum_{j=1}^{m} \frac{\partial V(x(t_k))}{\partial x} g_j\big(x(t_k)\big) h_j\big(x(t_k)\big). \quad (5.46)$$

It follows from the above inequality and condition of Eq. 5.5 that:

$$\frac{\partial V(x(t_k))}{\partial x}\left(f\big(x(t_k)\big) + \sum_{j=1}^{m} g_j\big(x(t_k)\big) u_{p,j}^{*,c}(t_k|t_k) \right)$$

$$\leq \frac{\partial V(x(t_k))}{\partial x}\left(f\big(x(t_k)\big) + \sum_{j=1}^{m} g_j\big(x(t_k)\big) h_j\big(x(t_k)\big) \right)$$

$$\leq -\alpha_3\big(\|x(t_k)\|\big). \quad (5.47)$$

Following the same approach as in the proof of Theorem 5.1, we know that if the condition of Eq. 5.25 is satisfied, then the state of the closed-loop system can be proved to be maintained in $\Omega_{\rho_{\min}}$ under the iterative DMPC architecture of Eqs. 5.38–5.44. ◻

Corollary 5.1 *Consider a class of linear time-invariant systems*:

$$\dot{x}(t) = Ax(t) + \sum_{i=1}^{m} B_i u_i(t), \quad (5.48)$$

where A and B_i are constant matrices with appropriate dimensions. If we define the inputs of the distributed LMPC of Eqs. 5.38–5.43 at iteration c as in Eq. 5.45, then at a sampling time t_k, as the iteration number $c \to \infty$, the optimal cost of the distributed optimization problem of Eqs. 5.38–5.43 converges to the optimal cost of the corresponding centralized control system.

Proof Taking into account that $x(t_k)$ and $h(x(t_k))$ are known at t_k, the constraint of Eq. 5.43 can be written in the following linear form:

$$C\big(x(t_k)\big) u_j(t_k) \leq D\big(x(t_k)\big), \quad (5.49)$$

where $C(x(t_k))$ and $D(x(t_k))$ are constants at each t_k and only depend on $x(t_k)$. This implies that the constraint of Eq. 5.43 is linear with respect to u_j. For a linear system, it is also easy to verify that the constraints of Eqs. 5.38–5.42 are convex. Therefore, the optimization problem of Eqs. 5.38–5.43 is convex. If the inputs of the distributed controllers at each iteration c are defined as in Eq. 5.45, then the convergence of the cost given by the distributed optimization problem of Eqs. 5.38–5.43 to the corresponding centralized control system can be proved following similar strategies used in [5, 98] for time t_k. ◻

Remark 5.7 Note that the DMPC designs have the same stability region Ω_ρ as the one of the nonlinear control law $h(x)$. When the stability of the nonlinear control law $h(x)$ is global (i.e., the stability region is the entire state space), then the stability of the DMPC designs is also global. Note also that for any initial condition in Ω_ρ, the DMPC designs are proved to be feasible.

Remark 5.8 The choice of the horizon of the DMPC designs does not affect the stability of the closed-loop system. For any horizon length $N \geq 1$, the closed-loop stability is guaranteed by the constraints of Eqs. 5.23 and 5.43. However, the choice of the horizon does affect the performance of the DMPC designs.

Remark 5.9 Note that because the manipulated inputs enter the dynamics of the system of Eq. 5.1 in an affine manner, the constraints designed in the LMPC optimization problems of Eqs. 5.17–5.23 and 5.38–5.43 to guarantee the closed-loop stability can be decoupled for different distributed controllers as in Eqs. 5.23 and 5.43.

Remark 5.10 In the sequential DMPC architecture presented in Sect. 5.4.1, the distributed controllers are evaluated in sequence, which implies that the minimal time to obtain a set of solutions to all the LMPCs is the sum of the evaluation times of all the LMPCs; whereas in the iterative DMPC architecture presented in Sect. 5.4.2, the distributed controllers are evaluated in parallel, which implies that the minimal time to obtain a set of solutions to all the LMPCs in each iteration is the largest evaluation time among all the LMPCs.

Remark 5.11 An alternative to the DMPC designs is to design a centralized MPC to compute all the inputs. A centralized LMPC design for the system of Eq. 5.1 based on the nonlinear control law $h(x)$ is as follows (please also see Sect. 2.6):

$$\min_{u_1 \ldots u_m \in S(\Delta)} \int_{t_k}^{t_k+N} \left[\left\| \tilde{x}(\tau) \right\|_{Q_c} + \sum_{i=1}^{m} \left\| u_i(\tau) \right\|_{R_{ci}} \right] d\tau, \tag{5.50}$$

$$\text{s.t.} \quad \dot{\tilde{x}}(t) = f\left(\tilde{x}(t)\right) + \sum_{i=1}^{m} g_i\left(\tilde{x}(t)\right) u_i, \tag{5.51}$$

$$u_i(t) \in U_i, \quad i = 1, \ldots, m, \tag{5.52}$$

$$\tilde{x}(t_k) = x(t_k), \tag{5.53}$$

$$\sum_{i=1}^{m} \frac{\partial V(x(t_k))}{\partial x} g_i\left(x(t_k)\right) u_i(t_k)$$

$$\leq \sum_{i=1}^{m} \frac{\partial V(x(t_k))}{\partial x} g_i\left(x(t_k)\right) h_i\left(x(t_k)\right), \tag{5.54}$$

where $\tilde{x}$ is the predicted trajectory of the nominal system with u_i, $i = 1, \ldots, m$, the input trajectory computed by this centralized LMPC. The optimal solution to this optimization problem is denoted by $u_{ci}^*(t|t_k)$, $i = 1, \ldots, m$, which is defined for $t \in [t_k, t_{k+N})$. The manipulated inputs of the closed-loop system of Eq. 5.1 under this centralized LMPC are defined as follows:

$$u_i(t) = u_{ci}^*(t|t_k), \quad i = 1, \ldots, m, \forall t \in [t_k, t_{k+1}). \tag{5.55}$$

In what follows, we refer to this controller as the centralized LMPC.

Remark 5.12 Note that the sequential (or iterative) DMPC is not a direct decomposition of the centralized LMPC because the set of constraints of Eq. 5.23 (or Eq. 5.43) for $j = 1, \ldots, m$ in the DMPC formulation of Eqs. 5.17–5.23 (or Eq. 5.38–5.43) imposes a different feasibility region from the one of the centralized LMPC of Eqs. 5.50–5.54 which has a single constraint (Eq. 5.54).

Remark 5.13 Note also that for general nonlinear systems, there is no guarantee that the closed-loop performance of one (centralized or distributed) MPC architecture discussed in this section should be superior than the others since the solutions provided by these MPC architectures are proved to be feasible and stabilizing but the superiority of the performance of one MPC architecture over another is not established. This is because the MPC designs are implemented in a receding horizon scheme and the prediction horizon is finite; and also because of the different MPC designs are not equivalent as we discussed in Remark 5.12 and the nonconvexity property as we discussed in Remark 5.5. In applications of these MPC architectures, especially for chemical process control in which nonconvex problems is a very common occurrence, simulations should be conducted before making decisions as to which architecture should be used.

5.4.3 *Application to an Alkylation of Benzene Process*

The process of alkylation of benzene with ethylene to produce ethylbenzene is widely used in the petrochemical industry. Dehydration of the product produces styrene, which is the precursor to polystyrene and many copolymers. Over the last two decades, several methods and simulation results of alkylation of benzene with catalysts have been reported in the literature. The process model developed in this section is based on these references [23, 44, 83, 117]. More specifically, the process considered in this work consists of four CSTRs and a flash tank separator, as shown in Fig. 5.3. The CSTR-1, CSTR-2 and CSTR-3 are in series and involve the alkylation of benzene with ethylene. Pure benzene is fed from stream F_1 and pure ethylene is fed from streams F_2, F_4 and F_6. Two catalytic reactions take place in CSTR-1, CSTR-2 and CSTR-3. Benzene (A) reacts with ethylene (B) and produces the desired product ethylbenzene (C) (reaction 1); ethylbenzene can further react with ethylene to form 1,3-diethylbenzene (D) (reaction 2) which is the byproduct. The effluent of CSTR-3, including the products and leftover reactants, is fed to a flash tank separator, in which most of benzene is separated overhead by vaporization

Fig. 5.3 Process flow diagram of alkylation of benzene

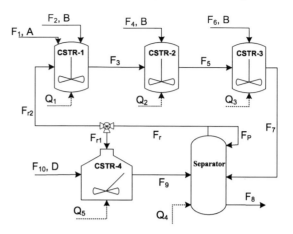

and condensation techniques and recycled back to the plant and the bottom product stream is removed. A portion of the recycle stream F_{r2} is fed back to CSTR-1 and another portion of the recycle stream F_{r1} is fed to CSTR-4 together with an additional feed stream F_{10} which contains 1,3-diethylbenzene from further distillation process that we do not consider in this example. In CSTR-4, reaction 2 and catalyzed transalkylation reaction in which 1,3-diethylbenzene reacts with benzene to produce ethylbenzene (reaction 3) takes place. All chemicals left from CSTR-4 eventually pass into the separator. All the materials in the reactions are in liquid phase due to high pressure and their molar volumes are assumed to be constants. The dynamic equations describing the behavior of the process, obtained through material and energy balances under standard modeling assumptions, are given below:

$$\frac{dC_{A1}}{dt} = \frac{F_1 C_{A0} + F_{r2} C_{Ar} - F_3 C_{A1}}{V_1} - r_1(T_1, C_{A1}, C_{B1}), \tag{5.56}$$

$$\frac{dC_{B1}}{dt} = \frac{F_2 C_{B0} + F_{r2} C_{Br} - F_3 C_{B1}}{V_1} - r_1(T_1, C_{A1}, C_{B1})$$
$$- r_2(T_1, C_{B1}, C_{C1}), \tag{5.57}$$

$$\frac{dC_{C1}}{dt} = \frac{F_{r2} C_{Cr} - F_3 C_{C1}}{V_1} + r_1(T_1, C_{A1}, C_{B1}) - r_2(T_1, C_{B1}, C_{C1}), \tag{5.58}$$

$$\frac{dC_{D1}}{dt} = \frac{F_{r2} C_{Dr} - F_3 C_{D1}}{V_1} + r_2(T_1, C_{B1}, C_{C1}), \tag{5.59}$$

$$\frac{dT_1}{dt} = \frac{Q_1 + F_1 C_{A0} H_A(T_{A0}) + F_2 C_{B0} H_B(T_{B0})}{\sum_i^{A,B,C,D} C_{i1} C_{pi} V_1}$$
$$+ \frac{\sum_i^{A,B,C,D} (F_{r2} C_{ir} H_i(T_4) - F_3 C_{i1} H_i(T_1))}{\sum_i^{A,B,C,D} C_{i1} C_{pi} V_1}$$
$$+ \frac{-\Delta H_{r1} r_1(T_1, C_{A1}, C_{B1}) - \Delta H_{r2} r_2(T_1, C_{B1}, C_{C1})}{\sum_i^{A,B,C,D} C_{i1} C_{pi}}, \tag{5.60}$$

$$\frac{dC_{A2}}{dt} = \frac{F_3 C_{A1} - F_5 C_{A2}}{V_2} - r_1(T_2, C_{A2}, C_{B2}), \tag{5.61}$$

$$\frac{dC_{B2}}{dt} = \frac{F_3 C_{B1} + F_4 C_{B0} - F_5 C_{B2}}{V_2} - r_1(T_2, C_{A2}, C_{B2})$$
$$- r_2(T_2, C_{B2}, C_{C2}), \tag{5.62}$$

$$\frac{dC_{C2}}{dt} = \frac{F_3 C_{C1} - F_5 C_{C2}}{V_2} + r_1(T_2, C_{A2}, C_{B2}) - r_2(T_2, C_{B2}, C_{C2}), \tag{5.63}$$

$$\frac{dC_{D2}}{dt} = \frac{F_3 C_{D1} - F_5 C_{R2}}{V_2} + r_2(T_2, C_{B2}, C_{C2}), \tag{5.64}$$

$$\frac{dT_2}{dt} = \frac{Q_2 + F_4 C_{B0} H_B(T_{B0})}{\sum_i^{A,B,C,D} C_{i2} C_{pi} V_2}$$
$$+ \frac{\sum_i^{A,B,C,D}(F_3 C_{i1} H_i(T_1) - F_5 C_{i2} H_i(T_2))}{\sum_i^{A,B,C,D} C_{i2} C_{pi} V_2}$$
$$+ \frac{-\Delta H_{r1} r_1(T_2, C_{A2}, C_{B2}) - \Delta H_{r2} r_2(T_2, C_{A2}, C_{B2})}{\sum_i^{A,B,C,D} C_{i2} C_{pi}}, \tag{5.65}$$

$$\frac{dC_{A3}}{dt} = \frac{F_5 C_{A2} - F_7 C_{A3}}{V_3} - r_1(T_3, C_{A3}, C_{B3}), \tag{5.66}$$

$$\frac{dC_{B3}}{dt} = \frac{F_5 C_{B2} + F_6 C_{B0} - F_7 C_{B3}}{V_3} - r_1(T_3, C_{A3}, C_{B3})$$
$$- r_2(T_3, C_{B3}, C_{C3}), \tag{5.67}$$

$$\frac{dC_{C3}}{dt} = \frac{F_5 C_{C2} - F_7 C_{C3}}{V_3} + r_1(T_3, C_{A3}, C_{B3}) - r_2(T_3, C_{B3}, C_{C3}), \tag{5.68}$$

$$\frac{dC_{D3}}{dt} = \frac{F_5 C_{D2} - F_7 C_{D3}}{V_3} + r_2(T_3, C_{B3}, C_{C3}), \tag{5.69}$$

$$\frac{dT_3}{dt} = \frac{Q_3 + F_6 C_{B0} H_B(T_{B0})}{\sum_i^{A,B,C,D} C_{i3} C_{pi} V_3}$$
$$+ \frac{\sum_i^{A,B,C,D}(F_5 C_{i2} H_i(T_2) - F_7 C_{i3} H_i(T_3))}{\sum_i^{A,B,C,D} C_{i3} C_{pi} V_3}$$
$$+ \frac{-\Delta H_{r1} r_1(T_3, C_{A3}, C_{B3}) - \Delta H_{r2} r_2(T_3, C_{B3}, C_{C3})}{\sum_i^{A,B,C,D} C_{i3} C_{pi}}, \tag{5.70}$$

$$\frac{dC_{A4}}{dt} = \frac{F_7 C_{A3} + F_9 C_{A5} - F_r C_{Ar} - F_8 C_{A4}}{V_4}, \tag{5.71}$$

$$\frac{dC_{B4}}{dt} = \frac{F_7 C_{B3} + F_9 C_{B5} - F_r C_{Br} - F_8 C_{B4}}{V_4}, \tag{5.72}$$

$$\frac{dC_{C4}}{dt} = \frac{F_7 C_{C3} + F_9 C_{C5} - F_r C_{Cr} - F_8 C_{C4}}{V_4}, \tag{5.73}$$

$$\frac{dC_{D4}}{dt} = \frac{F_7 C_{D3} + F_9 C_{D5} - F_r C_{Dr} - F_8 C_{D4}}{V_4}, \tag{5.74}$$

$$\frac{dT_4}{dt} = \frac{Q_4 + \sum_i^{A,B,C,D}(F_7 C_{i3} H_i(T_3) + F_9 C_{i5} H_i(T_5))}{\sum_i^{A,B,C,D} C_{i4} C_{pi} V_4}$$
$$+ \frac{\sum_i^{A,B,C,D}(-M_i H_i(T_4) - F_8 C_{i4} H_i(T_4) - M_i H_{vapi})}{\sum_i^{A,B,C,D} C_{i4} C_{pi} V_4}, \tag{5.75}$$

$$\frac{dC_{A5}}{dt} = \frac{F_{r1} C_{Ar} - F_9 C_{A5}}{V_5} - r_3(T_5, C_{A5}, C_{D5}), \tag{5.76}$$

$$\frac{dC_{B5}}{dt} = \frac{F_{r1} C_{Br} - F_9 C_{B5}}{V_5} - r_2(T_5, C_{B5}, C_{C5}), \tag{5.77}$$

$$\frac{dC_{C5}}{dt} = \frac{F_{r1} C_{Cr} - F_9 C_{C5}}{V_5} - r_2(T_5, C_{B5}, C_{C5})$$
$$+ 2r_3(T_5, C_{A5}, C_{D5}), \tag{5.78}$$

$$\frac{dC_{D5}}{dt} = \frac{F_{r1} C_{Dr} + F_{10} C_{D0} - F_9 C_{D5}}{V_5}$$
$$+ r_2(T_5, C_{B5}, C_{C5}) - r_3(T_5, C_{A5}, C_{D5}), \tag{5.79}$$

$$\frac{dT_5}{dt} = \frac{Q_5 + F_{10} C_{D0} H_D(T_{D0})}{\sum_i^{A,B,C,D} C_{i5} C_{pi} V_5}$$
$$+ \frac{\sum_i^{A,B,C,D}(F_{r1} C_{ir} H_i(T_4) - F_9 C_{i5} H_i(T_5))}{\sum_i^{A,B,C,D} C_{i5} C_{pi} V_5}$$
$$+ \frac{-\Delta H_{r2} r_2(T_5, C_{B5}, C_{C5}) - \Delta H_{r3} r_3(T_5, C_{A5}, C_{D5})}{\sum_i^{A,B,C,D} C_{i5} C_{pi}}, \tag{5.80}$$

where r_1, r_2 and r_3 are the reaction rates of reactions 1, 2 and 3, respectively and $H_i, i = A, B, C, D$, are the enthalpies of the reactants. The reaction rates are related to the concentrations of the reactants and the temperature in each reactor as follows:

$$r_1(T, C_A, C_B) = 0.0840 e^{\frac{-9502}{RT}} C_A^{0.32} C_B^{1.5}, \tag{5.81}$$

$$r_2(T, C_B, C_C) = \frac{0.0850 e^{\frac{-20643}{RT}} C_B^{2.5} C_C^{0.5}}{(1 + k_{EB2} C_D)}, \tag{5.82}$$

$$r_3(T, C_A, C_D) = \frac{66.1 e^{\frac{-61280}{RT}} C_A^{1.0218} C_D}{(1 + k_{EB3} C_A)}, \tag{5.83}$$

where:

$$k_{EB2} = 0.152e^{\frac{-3933}{RT}},\tag{5.84}$$

$$k_{EB3} = 0.490e^{\frac{-50870}{RT}}.\tag{5.85}$$

The heat capacities of the species are assumed to be constants and the molar enthalpies have a linear dependence on temperature as follows:

$$H_i(T) = H_{iref} + C_{pi}(T - T_{ref}), \quad i = A, B, C, D,\tag{5.86}$$

where C_{pi}, $i = A, B, C, D$ are heat capacities.

The model of the flash tank separator is developed under the assumption that the relative volatility of each species has a linear correlation with the temperature of the vessel within the operating temperature range of the flash tank, as shown below:

$$\alpha_A = 0.0449T_4 + 10,\tag{5.87}$$

$$\alpha_B = 0.0260T_4 + 10,\tag{5.88}$$

$$\alpha_C = 0.0065T_4 + 0.5,\tag{5.89}$$

$$\alpha_D = 0.0058T_4 + 0.25,\tag{5.90}$$

where α_i, $i = A, B, C, D$, represent the relative volatilities. It has also been assumed that there is a negligible amount of reaction taking place in the separator and a fraction of the total condensed overhead flow is recycled back to the reactors. The following algebraic equations model the composition of the overhead stream relative to the composition of the liquid holdup in the flash tank:

$$M_i = k\frac{\alpha_i(F_7C_{i3} + F_9C_{i5})\sum_j^{A,B,C,D}(F_7C_{j3} + F_9C_{j5})}{\sum_j^{A,B,C,D}\alpha_j(F_7C_{j3} + F_9C_{j5})}, \quad i = A, B, C, D,\tag{5.91}$$

where M_i, $i = A, B, C, D$ are the molar flow rates of the overhead reactants and k is the fraction of condensed overhead flow recycled to the reactors. Based on M_i, $i = A, B, C, D$, we can calculate the concentration of the reactants in the recycle streams as follows:

$$C_{ir} = \frac{M_i}{\sum_j^{A,B,C,D} M_i/C_{j0}}, \quad i = A, B, C, D,\tag{5.92}$$

where C_{j0}, $j = A, B, C, D$, are the mole densities of pure reactants. The condensation of vapor takes place overhead, and a portion of the condensed liquid is purged back to separator to keep the flow rate of the recycle stream at a fixed value. The temperature of the condensed liquid is assumed to be the same as the temperature of the vessel.

The definitions for the variables used in the above model can be found in Table 5.1, with the parameter values given in Table 5.2.

Table 5.1 Process variables of the alkylation of benzene process of Eqs. 5.56–5.80

$C_{A1}, C_{B1}, C_{C1}, C_{D1}$	Concentrations of A, B, C, D in CSTR-1
$C_{A2}, C_{B2}, C_{C2}, C_{D2}$	Concentrations of A, B, C, D in CSTR-2
$C_{A3}, C_{B3}, C_{C3}, C_{D3}$	Concentrations of A, B, C, D in CSTR-3
$C_{A4}, C_{B4}, C_{C4}, C_{D4}$	Concentrations of A, B, C, D in separator
$C_{A5}, C_{B5}, C_{C5}, C_{D5}$	Concentrations of A, B, C, D in CSTR-4
$C_{Ar}, C_{Br}, C_{Cr}, C_{Dr}$	Concentrations of A, B, C, D in F_r, F_{r1}, F_{r2}
T_1, T_2, T_3, T_4, T_5	Temperatures in each vessel
T_{ref}	Reference temperature
F_3, F_5, F_7, F_8, F_9	Effluent flow rates from each vessel
$F_1, F_2, F_4, F_6, F_{10}$	Feed flow rates to each vessel
F_r, F_{r1}, F_{r2}	Recycle flow rates
$H_{vapA}, H_{vapB}, H_{vapC}, H_{vapD}$	Enthalpies of vaporization of A, B, C, D
$H_{Aref}, H_{Bref}, H_{Cref}, H_{Dref}$	Enthalpies of A, B, C, D at T_{ref}
$\Delta H_{r1}, \Delta H_{r2}, \Delta H_{r3}$	Heat of reactions 1, 2 and 3
V_1, V_2, V_3, V_4, V_5	Volume of each vessel
Q_1, Q_2, Q_3, Q_4, Q_5	External heat/coolant inputs to each vessel
$C_{pA}, C_{pB}, C_{pC}, C_{pD}$	Heat capacity of A, B, C, D at liquid phase
$\alpha_A, \alpha_B, \alpha_C, \alpha_D$	Relative volatilities of A, B, C, D
$C_{A0}, C_{B0}, C_{C0}, C_{D0}$	Molar densities of pure A, B, C, D
T_{A0}, T_{B0}, T_{D0}	Feed temperatures of pure A, B, D
k	Fraction of overhead flow recycled to the reactors

Each of the tanks has an external heat/coolant input. The manipulated inputs to the process are the heat injected to or removed from the five vessels, Q_1, Q_2, Q_3, Q_4 and Q_5, and the feed stream flow rates to CSTR-2 and CSTR-3, F_4 and F_6.

The states of the process consist of the concentrations of A, B, C, D in each of the five vessels and the temperatures of the vessels. The state of the process is assumed to be available continuously to the controllers. We consider a stable steady state (operating point), x_s, of the process which is defined by the steady-state inputs Q_{1s}, Q_{2s}, Q_{3s}, Q_{4s}, Q_{5s}, F_{4s} and F_{6s} which are shown in Table 5.3 with corresponding steady-state values shown in Table 5.4.

The control objective is to regulate the system from an initial state to the steady state. The initial state values are shown in Table 5.6.

The first distributed controller (LMPC 1) will be designed to decide the values of Q_1, Q_2 and Q_3, the second distributed controller (LMPC 2) will be designed to decide the values of Q_4 and Q_5, and the third distributed controller (LMPC 3) will be designed to decide the values of F_4 and F_6. Taking this into account, the process model of Eqs. 5.56–5.80 belongs to the following class of nonlinear systems:

$$\dot{x}(t) = f(x) + g_1(x)u_1(t) + g_2(x)u_2(t) + g_3(x)u_3(t), \tag{5.93}$$

Table 5.2 Parameter values of the alkylation of benzene process of Eqs. 5.56–5.80

F_1	7.1×10^{-3} [m³/s]	F_r	0.012 [m³/s]
F_2	8.697×10^{-4} [m³/s]	F_{r1}	0.006 [m³/s]
F_{r2}	0.006 [m³/s]	V_1	1 [m³]
F_{10}	2.31×10^{-3} [m³/s]	V_2	1 [m³]
H_{vapA}	3.073×10^4 [J/mole]	V_3	1 [m³]
H_{vapB}	1.35×10^4 [J/mole]	V_4	3 [m³]
H_{vapC}	4.226×10^4 [J/mole]	V_5	1 [m³]
H_{vapD}	4.55×10^4 [J/mole]	C_{pA}	184.6 [J/mole K]
H_{Aref}	7.44×10^4 [J/mole]	H_{Bref}	5.91×10^4 [J/mole]
H_{Cref}	2.02×10^4 [J/mole]	H_{Bref}	-2.89×10^4 [J/mole]
ΔH_{r1}	-1.536×10^5 [J/mole]	C_{pB}	59.1 [J/mole K]
ΔH_{r2}	-1.118×10^5 [J/mole]	C_{pC}	247 [J/mole K]
ΔH_{r3}	4.141×10^5 [J/mole]	C_{pD}	301.3 [J/mole K]
C_{A0}	1.126×10^4 [mole/m³]	T_{ref}	450 [K]
C_{B0}	2.028×10^4 [mole/m³]	T_{A0}	473 [K]
C_{C0}	8174 [mole/m³]	T_{B0}	473 [K]
C_{D0}	6485 [mole/m³]	T_{D0}	473 [K]
k	0.8		

Table 5.3 Steady-state input values for x_s of the alkylation of benzene process of Eqs. 5.56–5.80

Q_{1s}	-4.4×10^6 [J/s]	Q_{2s}	-4.6×10^6 [J/s]
Q_{3s}	-4.7×10^6 [J/s]	Q_{4s}	9.2×10^6 [J/s]
Q_{5s}	5.9×10^6 [J/s]	F_{4s}	8.697×10^{-4} [m³/s]
F_{4s}	8.697×10^{-4} [m³/s]		

where the state x is the deviation of the state of the process from the steady state, $u_1^T = [u_{11} \ u_{12} \ u_{13}] = [Q_1 - Q_{1s} \ Q_2 - Q_{2s} \ Q_3 - Q_{3s}]$, $u_2^T = [u_{21} \ u_{22}] = [Q_4 - Q_{4s} \ Q_5 - Q_{5s}]$ and $u_3^T = [u_{31} \ u_{32}] = [F_4 - F_{4s} \ F_6 - F_{6s}]$ are the manipulated inputs which are subject to the constraints shown in Table 5.5.

In the control of the process, u_1 and u_2 are necessary to keep the stability of the closed-loop system, while u_3 can be used as an extra manipulated input to improve the closed-loop performance. To illustrate the theoretical results, we first design the nonlinear control law $h(x) = [h_1(x) \ h_2(x) \ h_3(x)]^T$. Specifically, $h_1(x)$ and $h_2(x)$ are designed as follows [97]:

$$h_i(x) = \begin{cases} -\dfrac{L_f V + \sqrt{(L_f V)^2 + (L_{g_i} V)^4}}{(L_{g_i} V)^2} L_{g_i} V & \text{if } L_{g_i} V \neq 0, \\ 0 & \text{if } L_{g_i} V = 0, \end{cases} \qquad (5.94)$$

where $i = 1, 2$, $L_f V = \frac{\partial V}{\partial x} f(x)$ and $L_{g_i} V = \frac{\partial V}{\partial x} g_i(x)$ denote the Lie derivatives of the scalar function V with respect to f and g_i ($i = 1, 2$), respectively. The controller $h_3(x)$ is chosen to be $h_3(x) = [0 \ 0]^T$ because the input set u_3 is not needed to

Table 5.4 Steady-state values for x_s of the alkylation of benzene process of Eqs. 5.56–5.80

C_{A1}	9.101×10^3 [mole/m^3]	C_{A2}	7.548×10^3 [mole/m^3]
C_{B1}	22.15 [mole/m^3]	C_{B2}	23.46 [mole/m^3]
C_{C1}	1.120×10^3 [mole/m^3]	C_{C2}	1.908×10^3 [mole/m^3]
C_{D1}	2.120×10^2 [mole/m^3]	C_{D2}	3.731×10^2 [mole/m^3]
T_1	4.772×10^2 [K]	T_2	4.77×10^2 [K]
C_{A3}	6.163×10^3 [mole/m^3]	C_{A4}	1.723×10^3 [mole/m^3]
C_{B3}	24.84 [mole/m^3]	C_{B4}	13.67 [mole/m^3]
C_{C3}	2.616×10^3 [mole/m^3]	C_{C4}	5.473×10^3 [mole/m^3]
C_{D3}	5.058×10^2 [mole/m^3]	C_{D4}	7.044×10^2 [mole/m^3]
T_3	4.735×10^2 [K]	T_4	4.706×10^2 [K]
C_{A5}	5.747×10^3 [mole/m^3]	C_{D5}	1.537×10^2 [mole/m^3]
C_{B5}	3.995 [mole/m^3]	T_5	4.783×10^2 [K]
C_{C5}	3.830×10^3 [mole/m^3]		

Table 5.5 Manipulated input constraints of the alkylation of benzene process of Eqs. 5.56–5.80

$|u_{11}| \leq 7.5 \times 10^5$ [J/s]	$|u_{12}| \leq 5 \times 10^5$ [J/s]
$|u_{13}| \leq 5 \times 10^5$ [J/s]	$|u_{21}| \leq 6 \times 10^5$ [J/s]
$|u_{22}| \leq 5 \times 10^5$ [J/s]	$|u_{31}| \leq 4.93 \times 10^{-5}$ [m^3/s]
$|u_{32}| \leq 4.93 \times 10^{-5}$ [m^3/s]	

Table 5.6 Initial state values of the alkylation of benzene process of Eqs. 5.56–5.80

C_{A1}	9.112×10^3 [mole/m^3]	C_{A2}	7.557×10^3 [mole/m^3]
C_{B1}	25.09 [mole/m^3]	C_{B2}	27.16 [mole/m^3]
C_{C1}	1.113×10^3 [mole/m^3]	C_{C2}	1.905×10^3 [mole/m^3]
C_{D1}	2.186×10^2 [mole/m^3]	C_{D2}	3.695×10^2 [mole/m^3]
T_1	4.430×10^2 [K]	T_2	4.371×10^2 [K]
C_{A3}	6.170×10^3 [mole/m^3]	C_{A4}	1.800×10^3 [mole/m^3]
C_{B3}	29.45 [mole/m^3]	C_{B4}	16.35 [mole/m^3]
C_{C3}	2.617×10^3 [mole/m^3]	C_{C4}	5.321×10^3 [mole/m^3]
C_{D3}	5.001×10^2 [mole/m^3]	C_{D4}	7.790×10^2 [mole/m^3]
T_3	4.284×10^2 [K]	T_4	4.331×10^2 [K]
C_{A5}	5.889×10^3 [mole/m^3]	C_{D5}	2.790×10^2 [mole/m^3]
C_{B5}	5.733 [mole/m^3]	T_5	4.576×10^2 [K]
C_{C5}	3.566×10^3 [mole/m^3]		

stabilize the process. We consider a Lyapunov function $V(x) = x^T P x$ with P being the following weight matrix:

$$P = diag\big([1\ 1\ 1\ 1\ 10\ 1\ 1\ 1\ 1\ 10\ 1\ 1\ 1\ 1\ 10\ 1\ 1\ 1\ 1\ 10\ 1\ 1\ 1\ 1\ 10]\big). \quad (5.95)$$

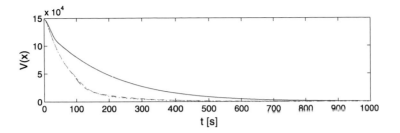

Fig. 5.4 Trajectories of the Lyapunov function $V(x)$ of the alkylation of benzene process of Eqs. 5.56–5.80 under the controller $h(x)$ of Eq. 5.94 implemented in a sample-and-hold fashion (*solid line*), the centralized LMPC of Eqs. 5.50–5.54 (*dashed line*), the sequential DMPC of Eqs. 5.17–5.23 (*dash-dotted line*) and the iterative DMPC of Eqs. 5.38–5.43 with $c = 1$ (*dotted line*)

The weights in P are chosen by a trial-and-error procedure. The basic idea behind this procedure is that more weight should be put on the temperatures of the five vessels because temperatures have more significant effect on the overall control performance, and the controller $h(x)$ should be able to stabilize the closed-loop system asymptotically with continuous feedback and actuation.

Based on $h(x)$, we design the centralized LMPC of Eqs. 5.50–5.54, the sequential DMPC of Eqs. 5.17–5.23 and the iterative DMPC of Eqs. 5.38–5.43. The sampling time used is $\Delta = 30$ s and the weight matrices:

$$Q_c = diag\left(\left[1\ 1\ 1\ 1\ 10^3\ 1\ 1\ 1\ 1\ 10^3\ 10\ 10\ 10\ 10\ 10^4\ 1\ 1\ 1\ 1\ 10^3\ 1\ 1\ 1\ 1\ 10^3\right]\right),\quad (5.96)$$

and $R_{c1} = diag([10^{-8}\ \ 10^{-8}\ \ 10^{-8}])$, $R_{c2} = diag([10^{-8}\ \ 10^{-8}])$ and $R_{c3} = diag([1\ 1])$.

First, we carried out a set of simulations which demonstrate that the nonlinear control law $h(x)$ and the different schemes of LMPCs can all stabilize the closed-loop system asymptotically. Figure 5.4 shows the trajectories of the Lyapunov function $V(x)$ under the different control schemes. Note that because of the constraints of Eqs. 5.54, 5.23 and 5.43, the trajectories of the Lyapunov function of the closed-loop system under the centralized LMPC, the sequential DMPC and the iterative DMPC are guaranteed to be bounded by the corresponding Lyapunov function trajectory under the controller $h(x)$ implemented in a sample-and-hold fashion with the sampling time Δ until $V(x)$ converges to a small region around the origin (i.e., $\Omega_{\rho_{min}}$). This point is also illustrated in Fig. 5.4.

Next, we compare the mean evaluation times of the centralized LMPC optimization problem and the sequential and iterative DMPC optimization problems. Each LMPC optimization problem was evaluated 100 times at different conditions. Different prediction horizons were considered in this set of simulations. The simulations were carried out using JAVA™ programming language in a PENTIUM® 3.20 GHz computer. The optimization problems were solved using the open source interior point optimizer Ipopt [109]. The results are shown in Table 5.7. From Table 5.7, we can see that in all cases, the time needed to solve the centralized LMPC

Table 5.7 Mean evaluation time of different LMPC optimization problems for 100 evaluations

Centralized LMPC		$N = 1$ (s)	$N = 3$ (s)	$N = 6$ (s)
		2.192	8.694	27.890
Sequential	LMPC 1	0.472	2.358	6.515
	LMPC 2	0.497	1.700	4.493
	LMPC 3	0.365	1.453	3.991
Iterative	LMPC 1	0.484	2.371	6.280
	LMPC 2	0.426	1.716	4.413
	LMPC 3	0.185	0.854	2.355

Table 5.8 Total performance costs along the closed-loop trajectories I of the alkylation of benzene process of Eqs. 5.56–5.80

				$J\ (\times 10^7)$				
Centralized				1.8858				
Sequential				1.8891				
c_{max}	1	3	5	7	9	11	13	15
Iterative	1.8955	1.8883	1.8867	1.8863	1.8862	1.8859	1.8858	1.8858

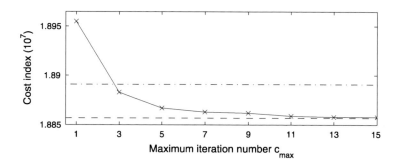

Fig. 5.5 Total performance costs along the closed-loop trajectories of the alkylation of benzene process of Eqs. 5.56–5.80 under centralized LMPC of Eqs. 5.50–5.54 (*dashed line*), sequential DMPC of Eqs. 5.17–5.23 (*dash-dotted line*) and iterative DMPC of Eqs. 5.38–5.43 (*solid line*)

is much larger than the time needed to solve the sequential or iterative DMPCs. This is because the centralized LMPC has to solve a much larger (in terms of decision variables) optimization problem than the DMPCs. We can also see that the evaluation time of the centralized LMPC is even larger than the sum of evaluation times of LMPC 1, LMPC 2 and LMPC 3 in the sequential DMPC, and the times needed to solve the DMPCs in both sequential and iterative distributed schemes are of the same order of magnitude.

In the following set of simulations, we compare the centralized LMPC and the two DMPC schemes from a performance index point of view. In this set of simulations, the prediction horizon is $N = 1$. To carry out this comparison, the same initial

Table 5.9 Total performance costs along the closed-loop trajectories II of the alkylation of benzene process of Eqs. 5.56–5.80

	$J\ (\times 10^7)$			
Centralized	5.052			
Sequential	7.039			
$c_{\max}$	1	3	5	6
Iterative	7.2286	7.2241	7.2240	7.2240

Table 5.10 Total performance costs along the closed-loop trajectories III of the alkylation of benzene process of Eqs. 5.56–5.80

	$J\ (\times 10^7)$		
Centralized	3.8564		
Sequential	3.6755		
$c_{\max}$	1	3	4
Iterative	3.6663	3.6639	3.6639

condition and parameters were used for the different control schemes and the total cost under each control scheme was computed as follows:

$$J = \int_{t_0}^{t_M} \left[\|x(\tau)\|_{Q_c} + \|u_1(\tau)\|_{R_{c1}} + \|u_2(\tau)\|_{R_{c2}} + \|u_3(\tau)\|_{R_{c3}} \right] d\tau, \qquad (5.97)$$

where $t_0 = 0$ is the initial time of the simulations and $t_M = 1000$ s is the end of the simulations. Table 5.8 shows the total cost along the closed-loop system trajectories (trajectories I) under the different control schemes. For the iterative DMPC design, different maximum number of iterations, $c_{\max}$, are used. From Table 5.8, we can see that in this set of simulations, the centralized LMPC gives the lowest performance cost, the sequential DMPC gives lower cost than the iterative DMPC when there is no iteration ($c_{\max} = 1$). However, as the iteration number c increases, the performance cost given by the iterative DMPC decreases and converges to the cost of the one corresponding to the centralized LMPC. This point is also shown in Fig. 5.5.

Note that the above set of simulations only represents one case of many possible cases. As we discussed in Remarks 5.5 and 5.13, there is no guaranteed convergence of the performance of distributed MPC to the performance of a centralized MPC and there is also no guaranteed superiority of the performance of one DMPC scheme over the others. In the following, we show two sets of simulations to illustrate these points. In both sets of simulations, we chose different matrices R_{c1} and R_{c2}, and all the other parameters (Q_c, R_{c3}, Δ, N) remained the same as the previous set of simulations. In the first set of simulations, we picked $R_{c1} = diag([5 \times 10^{-5}\ 5 \times 10^{-5}\ 5 \times 10^{-5}])$, $R_{c2} = diag([5 \times 10^{-5}\ 5 \times 10^{-5}])$. The total performance cost along the closed-loop system trajectories (trajectories II) under this simulation setting are shown in Table 5.9. From Table 5.9, we can see that the centralized LMPC provides a much lower cost than both the sequential and iterative distributed LMPCs. We can also see that as the number of iterations increases, the iterative distributed LMPC converges to a value which is different from the one obtained by the centralized LMPC. In the second set of simulations, we picked $R_{c1} = diag([1 \times 10^{-4}\ 1 \times 10^{-4}\ 1 \times 10^{-4}])$, $R_{c2} = diag([1 \times 10^{-4}\ 1 \times 10^{-4}])$

Fig. 5.6 Sequential DMPC for nonlinear systems subject to asynchronous measurements

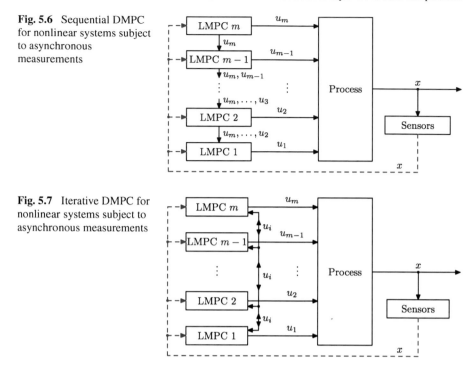

Fig. 5.7 Iterative DMPC for nonlinear systems subject to asynchronous measurements

and the total performance cost along the closed-loop system trajectories (trajectories III) are shown in Table 5.10 from which we can see that the centralized LMPC provides a higher cost than both distributed LMPCs.

5.5 Sequential and Iterative DMPC Designs with Asynchronous Measurements

In this section, we design sequential and iterative DMPC schemes, taking into account asynchronous measurements explicitly in their designs, that provide deterministic closed-loop stability properties. Similarly, in each DMPC architecture, we will design m LMPCs to compute u_i, $i = 1, \ldots, m$, and refer to the LMPC computing the input trajectories of u_i as LMPC i. Schematic diagrams of the sequential and iterative DMPC designs for systems subject to asynchronous measurements are shown in Figs. 5.6 and 5.7.

5.5.1 Modeling of Asynchronous Measurements

We assume that the state of the system of Eq. 5.1, $x(t)$, is available asynchronously at time instants t_a where $\{t_{a \geq 0}\}$ is a random increasing sequence of times. We also assume that there exists an upper bound T_m on the interval between two successive measurements, that is, the sequence satisfies the condition of Eq. 2.22.

5.5.2 Sequential DMPC with Asynchronous Measurements

5.5.2.1 Sequential DMPC Formulation

For the design of the sequential DMPC for systems subject to asynchronous measurements (see Fig. 5.6), we take advantage of the MPC scheme when feedback is lost to update the control inputs based on a state prediction obtained by the model and to have the control actuators store and implement the last computed optimal input trajectories. Specifically, the implementation strategy is as follows:

1. When a new measurement is available at t_a, all the LMPCs receive the state measurement $x(t_a)$ from the sensors.
2. For $j = m$ to 1
    2.1. LMPC j receives the entire future input trajectories of $u_i, i = m, \ldots, j+1$, from LMPC $j+1$ and evaluates the future input trajectory of u_j based on $x(t_a)$ and the received future input trajectories.
    2.2. LMPC j sends the entire input trajectories of u_j to its actuators and the entire input trajectories of $u_i, i = m, \ldots, j$, to LMPC $j-1$.
3. When a new measurement is received ($a \leftarrow a + 1$), go to Step 1.

In order to make a decision, LMPC j, $j = m, \ldots, 2$ must assume trajectories for $u_i, i = j - 1, \ldots, 1$, along the prediction horizon since the communication is one-directional. To this end, the controller $h(x)$ is used. In order to inherit the stability properties of the controller $h(x)$, each control input $u_i, i = 1, \ldots, m$ must satisfy a set of constraints that guarantee a given minimum contribution to the decrease rate of the Lyapunov function $V(x)$ in the case of asynchronous measurements. To this end, the input trajectories, $u_{n,i}(t|t_a)$ ($i = 1, \ldots, m$), defined in Eq. 5.37 are used.

Specifically, the design of LMPC j, $j = 1, \ldots, m$, is based on the following optimization problem:

$$\min_{u_j \in S(\Delta)} \int_{t_a}^{t_a+N\Delta} \left[\left\| \tilde{x}^j(\tau) \right\|_{Q_c} + \sum_{i=1}^{m} \left\| u_i(\tau) \right\|_{R_{ci}} \right] d\tau, \tag{5.98}$$

$$\text{s.t.} \quad \dot{\tilde{x}}^j(t) = f\left(\tilde{x}^j(t)\right) + \sum_{i=1}^{m} g_i\left(\tilde{x}^j(t)\right) u_i(t), \tag{5.99}$$

$$\dot{\tilde{x}}^j(t) = f\left(\hat{x}^j(t)\right) + \sum_{i=1}^{j} g_i\left(\hat{x}^j(t)\right) u_{n,i}(t|t_a) + \sum_{i=j+1}^{m} g_i\left(\hat{x}^j(t)\right) u_i(t), \tag{5.100}$$

$$u_i(t) = u_{n,i}(t|t_a), \quad i = 1, \ldots, j-1, \tag{5.101}$$

$$u_i(t) = u_{s,i}^{a,*}(t|t_a), \quad i = j+1, \ldots, m, \tag{5.102}$$

$$u_{s,j}(t) \in U_j, \tag{5.103}$$

$$\tilde{x}^j(t_a) = \hat{x}^j(t_a) = x(t_a), \tag{5.104}$$

$$V(\tilde{x}^j(t)) \leq V(\hat{x}^j(t)), \quad \forall t \in [t_a, t_a + N_R \Delta), \tag{5.105}$$

where N_R is the smallest integer satisfying $T_m \leq N_R \Delta$. The vector $\tilde{x}^j$ is the predicted trajectory of the nominal system with u_j computed by the above optimization problem (i.e., LMPC j) and the other control inputs defined by Eqs. 5.101–5.102. The vector $\hat{x}^j$ is the predicted trajectory of the nominal system with $u_j = u_{n,j}(t|t_a)$ and the other control inputs defined by Eqs. 5.101–5.102. In order to fully take advantage of the prediction, we choose $N \geq N_R$. The optimal solution to this optimization problem is denoted $u_{s,j}^{a,*}(t|t_a)$ and is defined for $t \in [t_a, t_a + N\Delta)$.

The constraint of Eq. 5.99 is the nominal model of the system, which is used to generate the trajectory $\tilde{x}^j$; the constraint of Eq. 5.100 defines a reference trajectory of the nominal system (i.e., $\hat{x}^j$) when the input u_j is defined by $u_{n,j}(t|t_a)$; the constraint of Eq. 5.101 defines the value of the inputs evaluated after u_j (i.e., u_i with $i = 1, \ldots, j-1$); the constraint of Eq. 5.102 defines the value of the inputs evaluated before u_j (i.e., u_i with $i = j+1, \ldots, m$); the constraint of Eq. 5.103 is the constraint on the manipulated input u_j; the constraint of Eq. 5.104 sets the initial state for the optimization problem; and the constraint of Eq. 5.105 guarantees that the contribution of input u_j to the decrease rate of the time derivative of the Lyapunov function from t_a to $t_a + N_R \Delta$, if $u_j = u_{s,j}^{a,*}(t|t_a)$, $t \in [t_a, t_a + N_R \Delta)$ is applied, is bigger or equal to the value obtained when $u_j = u_{n,j}(t|t_a)$, $t \in [t_a, t_a + N_R \Delta)$ is applied. This constraint guarantees that the sequential DMPC design of Eqs. 5.98–5.105 maintains the stability of the nonlinear control law $h(x)$ implemented in a sample-and-hold fashion and with open-loop state estimation in the presence of asynchronous measurements.

The manipulated inputs of the closed-loop system under the above sequential DMPC are defined as follows:

$$u_i(t) = u_{s,i}^{a,*}(t|t_k), \quad i = 1, \ldots, m, \forall t \in [t_a, t_{a+1}). \tag{5.106}$$

5.5.2.2 Stability Properties

The sequential DMPC design of Eqs. 5.98–5.105 maintains the closed-loop stability properties of the nonlinear control law $h(x)$ implemented in a sample-and-hold fashion and with open-loop state estimation in the presence of asynchronous measurements. This property is presented in Theorem 5.3 below. To state this theorem, we need the following corollaries.

From Proposition 2.1, we can obtain the following corollary for systems with m sets of control inputs, entering the dynamics of the system in an affine fashion, which ensures that if the nominal system of Eq. 5.1 under the control $u_i = h_i(x)$

$(i = 1, \ldots, m)$ implemented in a sample-and-hold fashion with state feedback every sampling time starts in Ω_ρ, then it is ultimately bounded in $\Omega_{\rho_{\min}}$.

Corollary 5.2 *Consider the nominal sampled trajectory $\hat{x}$ of the system of Eq. 5.1 in closed-loop with a nonlinear control law $u_i = h_i(x)$ ($i = 1, \ldots, m$), satisfying the conditions of Eqs. 5.5–5.8 and applied in a sample-and-hold fashion, obtained by solving recursively the following equation:*

$$\dot{\hat{x}}(t) = f\big(\hat{x}(t)\big) + \sum_{i=1}^{m} g_i\big(\hat{x}(t)\big) h_i\big(\hat{x}(t_a)\big), \quad t \in [t_k, t_{k+1}), \tag{5.107}$$

where $t_k = t_0 + k\Delta$, $k = 0, 1, \ldots$. Let $\Delta, \varepsilon_s > 0$ and $\rho > \rho_s > 0$ satisfy:

$$-\alpha_3\big(\alpha_2^{-1}(\rho_s)\big) + L'M \le -\varepsilon_s/\Delta \tag{5.108}$$

with $L' = L'_x + \sum_{i=1}^{m} L'_{u_i} u_i^{\max}$. Then, if $\rho_{\min} < \rho$ where $\rho_{\min}$ is defined as in Eq. 5.26 and $\hat{x}(0) \in \Omega_\rho$, the following inequality holds:

$$V\big(\hat{x}(t)\big) \le V\big(\hat{x}(t_k)\big), \quad \forall t \in [t_k, t_{k+1}), \tag{5.109}$$

$$V\big(\hat{x}(t_k)\big) \le \max\big\{ V\big(\hat{x}(t_0)\big) - k\varepsilon_s, \rho_{\min} \big\}. \tag{5.110}$$

Proof Following the definition of $\hat{x}(t)$ in Eq. 5.107, the time derivative of the Lyapunov function $V(x)$ along the trajectory $\hat{x}(t)$ of the system 5.1 in $t \in [t_k, t_{k+1})$ is given by:

$$\dot{V}\big(\hat{x}(t)\big) = \frac{\partial V(\hat{x}(t))}{\partial x}\left(f\big(\hat{x}(t)\big) + \sum_{i=1}^{m} g_i\big(\hat{x}(t)\big) h_i\big(\hat{x}(t_k)\big) \right). \tag{5.111}$$

Adding and subtracting $\frac{\partial V(\hat{x}(t_k))}{\partial x}\big(f(x(t_k)) + \sum_{i=1}^{m} g_i(x(t_k)) h_i(\hat{x}(t_k))\big)$ and taking into account Eq. 5.6, we obtain:

$$\dot{V}\big(\hat{x}(t)\big) \le -\alpha_3\big(\|\hat{x}(t_k)\|\big) + \frac{\partial V(\hat{x}(t))}{\partial x}\left(f\big(\hat{x}(t)\big) + \sum_{i=1}^{m} g_i\big(\hat{x}(t)\big) h_i\big(\hat{x}(t_k)\big) \right)$$

$$- \frac{\partial V(\hat{x}(t_k))}{\partial x}\left(f\big(x(t_k)\big) + \sum_{i=1}^{m} g_i\big(x(t_k)\big) h_i\big(\hat{x}(t_k)\big) \right). \tag{5.112}$$

From the Lipschitz property of Eqs. 5.14–5.15, the fact that the control inputs are bounded in convex sets and the above inequality of Eq. 5.112, we have that:

$$\dot{V}\big(\hat{x}(t)\big) \le -\alpha_3\big(\alpha_2^{-1}(\rho_s)\big) + \left(L'_x + \sum_{i=1}^{m} L'_{u_i} u_i^{\max} \right) \|\hat{x}(t) - \hat{x}(t_k)\| \tag{5.113}$$

for all $\hat{x}(t_k) \in \Omega_\rho / \Omega_{\rho_s}$. Taking into account the Lipschitz property of Eq. 5.9 and the continuity of $\hat{x}(t)$, the following bound can be written for all $t \in [t_k, t_{k+1})$:

$$\left\| \hat{x}(t) - \hat{x}(t_k) \right\| \le M\Delta. \tag{5.114}$$

Using the expression of Eq. 5.114, we obtain the following bound on the time derivative of the Lyapunov function for $t \in [t_k, t_{k+1})$, for all initial states $\hat{x}(t_k) \in \Omega_\rho / \Omega_{\rho_s}$:

$$\dot{V}\left(\hat{x}(t)\right) \le -\alpha_3\left(\alpha_2^{-1}(\rho_s)\right) + L'M\Delta, \tag{5.115}$$

where $L' = L'_x + \sum_{i=1}^{m} L'_{u_i} u_i^{\max}$. If the condition of Eq. 5.108 is satisfied, then $\dot{V}(\hat{x}(t)) \le -\varepsilon_s / \Delta$. Integrating this bound on $t \in [t_k, t_{k+1})$ we obtain that the inequality of Eq. 5.109 holds. Using Eq. 5.109 recursively, it is proved that, if $x(t_0) \in \Omega_\rho / \Omega_{\rho_s}$, the state converges to Ω_{ρ_s} in a finite number of sampling times without leaving the stability region. Once the state converges to $\Omega_{\rho_s} \subseteq \Omega_{\rho_{\min}}$, it remains inside $\Omega_{\rho_{\min}}$ for all times. This statement holds because of the definition of $\rho_{\min}$ as in Eq. 5.26. ◻

From Proposition 2.2, we can have the following Corollary 5.3 to get an upper bound on the deviation of the state trajectory obtained using the nominal model of Eq. 5.1, from the real-state trajectory when the same control actions are applied for systems with m sets of control inputs entering the dynamics of the system in an affine fashion.

Corollary 5.3 *Consider the systems*:

$$\dot{x}_a(t) = f\left(x_a(t)\right) + \sum_{i=1}^{m} g_i\left(x_a(t)\right) u_i(t) + k\left(x_a(t)\right) w(t), \tag{5.116}$$

$$\dot{x}_b(t) = f\left(x_b(t)\right) + \sum_{i=1}^{m} g_i\left(x_b(t)\right) u_i(t), \tag{5.117}$$

where initial states $x_a(t_0), x_b(t_0) \in \Omega_\rho$ *with* $x_b(t_0) = x_a(t_0) + n_x$ *and* $\|n_x\| \le \theta_x$. *There exists a function* $f_W(\cdot, \cdot)$ *such that*:

$$\left\| x_a(t) - x_b(t) \right\| \le f_W(\theta_x, t - t_0), \tag{5.118}$$

for all $x_a(t), x_b(t) \in \Omega_\rho$ *and all* $w(t) \in W$ *with*:

$$f_W(\theta_x, \tau) = \left(\frac{L_w \theta}{L''} + \theta_x \right) \left(e^{L''\tau} - 1 \right), \tag{5.119}$$

where $L'' = L_x + \sum_{i=1}^{m} L_{u_i} u_i^{\max}$.

Proof Define the error vector as $e(t) = x_a(t) - x_b(t)$. The time derivative of the error is given by:

$$\dot{e}(t) = f\big(x_a(t)\big) - f\big(x_b(x(t))\big) + \sum_{i=1}^{m}\big(g_i\big(x_a(t)\big) - g_i\big(x_b(t)\big)\big)u_i(t) + k\big(x_a(t)\big)w(t).$$

(5.120)

From the Lipschitz property of Eq. 5.11–5.13 and the fact that the control inputs are bounded in convex sets, the following inequality holds:

$$\big\|\dot{e}(t)\big\| \le L_x\big\|x_a(t) - x_b(t)\big\| + \sum_{i=1}^{m} L_{u_i}\big\|x_a(t) - x_b(t)\big\|\big\|L_w\big\|\big\|w(t)\big\|$$

$$\le \left(L_x + \sum_{i=1}^{m} L_{u_i}\right)u_i^{\max}\big\|e(t)\big\| + L_w\theta$$

(5.121)

for all $x_a(t), x_b(t) \in \Omega_\rho$ and $w(t) \in W$. Integrating $\|\dot{e}(t)\|$ with initial condition $\|e(t_0)\| = \|n_x\|$ and that $\|n_x\| \le \theta_x$, the following bound on the norm of the error vector is obtained:

$$\big\|e(t)\big\| \le \left(\frac{L_w\theta}{L''} + \theta_x\right)\big(e^{L''(t-t_0)} - 1\big),$$

(5.122)

where $L'' = L_x + \sum_{i=1}^{m} L_{u_i}u_i^{\max}$. This implies that the inequality of Eq. 5.118 holds for:

$$f_W(\tau) = \left(\frac{L_w\theta}{L''} + \theta_x\right)\big(e^{L''\tau} - 1\big),$$

(5.123)

which proves this corollary. □

In Theorem 5.3 below, we provide sufficient conditions under which the DMPC of Eqs. 5.98–5.106 guarantees that the state of the closed-loop system is ultimately bounded in a region that contains the origin.

Theorem 5.3 *Consider the system of Eq. 5.1 in closed-loop with x available at asynchronous sampling time instants $\{t_{a\ge0}\}$, satisfying the condition of Eq. 2.22, under the DMPC design of Eqs. 5.98–5.106 based on a control law h(x) that satisfies the conditions of Eqs. 4.3–5.8. Let $\Delta, \varepsilon_s > 0$, $\rho > \rho_{\min} > 0$, $\rho > \rho_s > 0$ and $N \ge N_R \ge 1$ satisfy the conditions of Eqs. 5.108 and the following inequality:*

$$-N_R\varepsilon_s + f_V\big(f_W(0, N_R\Delta)\big) < 0$$

(5.124)

with f_V defined in Eq. 2.49 and f_W defined in Eq. 5.119, and N_R being the smallest integer satisfying $N_R\Delta \ge T_m$. If the initial state of the closed-loop system $x(t_0) \in \Omega_\rho$, then x(t) is ultimately bounded in $\Omega_{\rho_a} \subseteq \Omega_\rho$ where:

$$\rho_a = \rho_{\min} + f_V\big(f_W(0, N_R\Delta)\big)$$

(5.125)

with $\rho_{\min}$ defined in Eq. 5.26.

Proof In order to prove that the state of the closed-loop system is ultimately bounded in a region that contains the origin, we prove that $V(x(t_a))$ is a decreasing sequence of values with a lower bound. Specifically, we focus on the time interval $t \in [t_a, t_{a+1})$ and prove that $V(x(t_{a+1}))$ is reduced compared with $V(x(t_a))$ or is maintained in an invariant set containing the origin.

To simplify the notation, we assume that all the signals used in this proof refer to the different optimization problems solved at t_a with the initial condition $x(t_a)$, and the trajectory $\tilde{x}^j(t)$, $j = 1, \ldots, m$, is corresponding to the optimal input $u_{s,j+1}^{a,*}(t|t_a)$. We also note that the predicted trajectories $\tilde{x}^{j+1}(t)$ and $\hat{x}^j(t)$ generated in the optimization problems of LMPC $j + 1$ and LMPC j are identical. This property will be used in the proof.

Part 1: In this part, we prove that the stability results stated in Theorem 5.3 hold in the case that $t_{a+1} - t_a = T_m$ for all a and $T_m = N_R \Delta$. This case corresponds to the worst situation in the sense that the controllers need to operate in open-loop for the maximum possible amount of time. By Corollary 5.2 and the fact that $t_{a+1} = t_a + N_R \Delta$, the following inequality is obtained:

$$V\big(\hat{x}(t_{a+1})\big) \leq \max\big\{V\big(\hat{x}(t_a)\big) - N_R \varepsilon_s, \rho_{\min}\big\}. \tag{5.126}$$

From the constraints of Eq. 5.105 in the LMPCs, the following inequality can be written:

$$V\big(\tilde{x}^j(t)\big) \leq V\big(\hat{x}^j(t)\big), \quad j = 1, \ldots, m, \forall t \in [t_a, t_a + N_R \Delta). \tag{5.127}$$

By the fact that $\tilde{x}^{j+1}(t)$ and $\hat{x}^j(t)$ are identical, the following equations can be written:

$$V\big(\hat{x}^j(t)\big) = V\big(\tilde{x}^{j+1}(t)\big), \quad j = 1, \ldots, m - 1, \forall t \in [t_a, t_a + N_R \Delta). \tag{5.128}$$

From the inequalities of Eqs. 5.127 and 5.128, the following inequalities are obtained:

$$V\big(\tilde{x}^1(t)\big) \leq \cdots \leq V\big(\tilde{x}^j(t)\big) \leq \cdots \leq V\big(\tilde{x}^m(t)\big) \leq V\big(\hat{x}^m(t)\big), \quad \forall t \in [t_a, t_a + N_R \Delta). \tag{5.129}$$

Note that the trajectory $\tilde{x}^1$ is the nominal trajectory (i.e., $\tilde{x}$) of the closed-loop system under the control of the sequential DMPC of Eqs. 5.98–5.106. Note also that the trajectory $\hat{x}^m$ is the nominal sampled trajectory (i.e., $\hat{x}$) of the closed-loop system defined in Eq. 5.107. Therefore, the following trajectory can be written:

$$V\big(\tilde{x}(t)\big) \leq V\big(\hat{x}(t)\big), \quad \forall t \in [t_a, t_a + N_R \Delta). \tag{5.130}$$

From the inequalities of Eq. 5.126 and 5.130 and the fact that $\hat{x}(t_a) = x(t_a)$, the following inequality is obtained:

$$V\big(\tilde{x}(t_{a+1})\big) \leq \max\big\{V\big(x(t_a)\big) - N_R \varepsilon_s, \rho_{\min}\big\}. \tag{5.131}$$

When $x(t) \in \Omega_\rho$ for all times (this point will be proved below), we can apply Proposition 2.3 to obtain the following inequality:

$$V\big(x(t_{a+1})\big) \leq V\big(\tilde{x}(t_{a+1})\big) + f_V\big(\|\tilde{x}(t_{a+1}) - x(t_{a+1})\|\big). \qquad (5.132)$$

Applying Corollary 5.3, we obtain the following upper bound on the deviation of $\tilde{x}(t)$ from $x(t)$:

$$\|x(t_{a+1}) - \tilde{x}(t_{a+1})\| \leq f_W(0, N_R \Delta). \qquad (5.133)$$

From the inequalities of Eqs. 5.132 and 5.133, the following upper bound on $V(x(t_{a+1}))$ can be written:

$$V\big(x(t_{a+1})\big) \leq V\big(\tilde{x}(t_{a+1})\big) + f_V\big(f_W(0, N_R \Delta)\big). \qquad (5.134)$$

Using the inequality of Eq. 5.131, we can rewrite the inequality of Eq. 5.134 as follows:

$$V\big(x(t_{a+1})\big) \leq \max\big\{V\big(x(t_a)\big) - N_R \varepsilon_s, \rho_{\min}\big\} + f_V\big(f_W(0, N_R \Delta)\big). \qquad (5.135)$$

If the condition of Eq. 5.124 is satisfied, from the inequality of Eq. 5.135, we know that there exists $\varepsilon_w > 0$ such that the following inequality holds:

$$V\big(x(t_{a+1})\big) \leq \max\big\{V\big(x(t_a)\big) - \varepsilon_w, \rho_a\big\} \qquad (5.136)$$

which implies that if $x(t_a) \in \Omega_\rho / \Omega_{\rho_a}$, then $V(x(t_{a+1})) < V(x(t_a))$, and if $x(t_a) \in \Omega_{\rho_a}$, then $V(x(t_{a+1})) \leq \rho_a$.

Because the upper bound on the difference between the Lyapunov function of the actual trajectory x and the nominal trajectory $\tilde{x}$ is a strictly increasing function of time (see Corollary 5.3 and Proposition 2.3 for the expressions of $f_V(\cdot)$ and $f_W(\cdot)$), the inequality of Eq. 5.136 also implies that

$$V\big(x(t)\big) \leq \max\big\{V\big(x(t_a)\big), \rho_a\big\}, \quad \forall t \in [t_a, t_{a+1}). \qquad (5.137)$$

Using the inequality of Eq. 5.137 recursively, it can be proved that if $x(t_0) \in \Omega_\rho$, then the closed-loop trajectories of the system of Eq. 5.1 under the sequential DMPC of Eqs. 5.98–5.106 stay in Ω_ρ for all times (i.e., $x(t) \in \Omega_\rho$, $\forall t$). Moreover, using the inequality of Eq. 5.137 recursively, it can be proved that if $x(t_0) \in \Omega_\rho$, the closed-loop trajectories of the system of Eq. 5.1 under the sequential DMPC of Eqs. 5.98–5.106 satisfy

$$\limsup_{t \to \infty} V\big(x(t)\big) \leq \rho_a. \qquad (5.138)$$

This proves that $x(t) \in \Omega_\rho$ for all times and $x(t)$ is ultimately bounded in Ω_{ρ_a} for the case when $t_{a+1} - t_a = T_m$ for all a and $T_m = N_R \Delta$.

Part 2: In this part, we extend the results proved in Part 1 to the general case, that is, $t_{a+1} - t_a \leq T_m$ for all a and $T_m \leq N_R \Delta$ which implies that $t_{a+1} - t_a \leq N_R \Delta$. Because $f_V(\cdot)$ and $f_W(\cdot)$ are strictly increasing functions of time and $f_V(\cdot)$ is convex, following similar steps as in Part 1, it can be shown that the inequality of Eq. 5.135 still holds. This proves that the stability results stated in Theorem 5.3 hold. $\square$

Remark 5.14 Note that the stability results stated in Theorem 5.3 also hold when the sequential DMPC of Eqs. 5.98–5.105 is applied to a nonlinear system described by Eq. 5.4.

5.5.3 Iterative DMPC with Asynchronous Measurements

5.5.3.1 Iterative DMPC Formulation

In contrast to the one-directional communication of the sequential DMPC architecture, the iterative DMPC architecture utilizes a bidirectional communication strategy in which all the distributed controllers are able to share their future input trajectories information after each iteration. In the presence of asynchronous measurements, the iterative DMPC of Eqs. 5.38–5.44 presented in Sect. 5.4.2 cannot guarantee closed-loop stability. In this subsection, we modify the implementation strategy and the formulation of the distributed controllers to take into account asynchronous measurements (see Fig. 5.7). The implementation strategy is as follows:

1. When a new measurement is available at t_a, all the LMPCs receive the state measurement $x(t_a)$ from the sensors and then evaluate their future input trajectories in an iterative fashion with initial input guesses generated by $h(\cdot)$.
2. At iteration c ($c \geq 1$):
    2.1. Each LMPC evaluates its own future input trajectory based on $x(t_a)$ and the latest received input trajectories of all the other LMPCs (when $c = 1$, initial input guesses generated by $h(\cdot)$ are used).
    2.2. The controllers exchange their future input trajectories. Based on all the input trajectories, each controller calculates and stores the value of the cost function.
3. If a termination condition is satisfied, each LMPC sends its entire future input trajectory corresponding to the smallest value of the cost function to its actuators; if the termination condition is not satisfied, go to Step 2 ($c \leftarrow c + 1$).
4. When a new measurement is received ($a \leftarrow a + 1$), go to Step 1.

The design of the LMPC j, $j = 1, \ldots, m$, at iteration c is based on the following optimization problem:

$$\min_{u_j \in S(\Delta)} \int_{t_a}^{t_a + N\Delta} \left[\left\| \tilde{x}^j(\tau) \right\|_{Q_c} + \sum_{i=1}^{m} \left\| u_i(\tau) \right\|_{R_{ci}} \right] d\tau, \tag{5.139}$$

$$\text{s.t.} \quad \dot{\tilde{x}}^j(t) = f\left(\tilde{x}^j(t)\right) + \sum_{i=1}^{m} g_i\left(\tilde{x}^j(t)\right) u_i, \tag{5.140}$$

$$u_i(t) = u_{p,i}^{a,c-1}(t|t_a), \quad \forall i \neq j, \tag{5.141}$$

$$\left\| u_j(t) - u_{p,j}^{a,c-1}(t|t_a) \right\| \leq \Delta u_j, \quad \forall t \in [t_a, t_a + N_R\Delta), \tag{5.142}$$

$$u_j(t) \in U_j, \tag{5.143}$$

$$\tilde{x}^j(t_a) = x(t_a), \tag{5.144}$$

$$\frac{\partial V(\tilde{x}^j(t))}{\partial \tilde{x}^j} \left(\frac{1}{m} f(\tilde{x}^j(t)) + g_j(\tilde{x}^j(t)) u_j(t) \right)$$

$$\leq \frac{\partial V(\hat{x}(t|t_a))}{\partial \hat{x}} \left(\frac{1}{m} f(\hat{x}(t|t_a)) + g_j(\hat{x}(t|t_a)) u_{n,j}(t|t_a) \right),$$

$$\forall t \in [t_a, t_a + N_R \Delta), \tag{5.145}$$

where $\tilde{x}^j$ is the predicted trajectory of the nominal system of Eq. 5.1 with u_j computed by this LMPC and all the other inputs are the optimal input trajectories at iteration $c - 1$ of the rest of the distributed controllers, $\hat{x}(t|t_a)$ and $u_{n,i}(t|t_a)$ ($i = 1, \ldots, m$) are defined in Eqs. 5.36 and 5.37, respectively. The optimal solution to this optimization problem is denoted $u_{p,j}^{a,c}(t|t_a)$ which is defined for $t \in [t_a, t_a + N\Delta)$. Accordingly, we define the final optimal input trajectory of LMPC j of Eqs. 5.139–5.145 as $u_{p,j}^{a,*}(t|t_a)$ which is also defined for $t \in [t_a, t_a + N\Delta)$.

Similar to the iterative DMPC with continuous measurements, for the first iteration of each distributed LMPC, the input trajectories defined in Eq. 5.37 based on the trajectory generated in Eq. 5.36 are used as the initial input trajectory guesses; that is, $u_{p,i}^{a,0} = u_{n,i}$ with $i = 1, \ldots, m$.

The constraint of Eq. 5.142 puts a limit on the input change in two consecutive iterations. This constraint allows LMPC j of Eqs. 5.139–5.145 to take advantage of the input trajectories received in the last iteration (i.e., $u_{p,i}^{a,c-1}$, $\forall i \neq j$) to predict the future evolution of the system state without introducing big errors. For LMPC j (i.e., u_j), the magnitude of input change in two consecutive iterations is restricted to be smaller than a positive constant Δu_j. Note that this constraint does not restrict the input to be in a small region and as the iteration number increases, the final optimal input could be quite different from the initial guess. The constraint of Eq. 5.145 is used to guarantee the closed-loop stability.

The manipulated inputs of the closed-loop system under the above iterative DMPC are defined as follows:

$$u_i(t) = u_{p,i}^{a,*}(t|t_a), \quad i = 1, \ldots, m, \ \forall t \in [t_a, t_{a+1}). \tag{5.146}$$

5.5.3.2 Stability Properties

The iterative DMPC design of Eqs. 5.139–5.146 takes into account asynchronous measurements explicitly in the controller design and the implementation strategy. It maintains the closed-loop stability properties of the nonlinear control law $h(x)$ implemented in a sample-and-hold fashion and with open-loop state estimation. This property is presented in Theorem 5.4. To state this theorem, we need another proposition.

Proposition 5.1 *Consider the systems:*

$$\dot{x}_a(t) = f\big(x_a(t)\big) + \sum_{i=1}^{m} g_i\big(x_a(t)\big)u_i^c(t), \qquad (5.147)$$

$$\dot{x}_b(t) = f\big(x_b(t)\big) + \sum_{i=1}^{m,\ i\neq j} g_i\big(x_b(t)\big)u_i^{c-1}(t) + g_j\big(x_b(t)\big)u_j^c(t) \qquad (5.148)$$

with initial states $x_a(t_0), x_b(t_0) \in \Omega_\rho$ such that $x_b(t_0) = x_a(t_0) + n_x$ and $\|n_x\| \leq \theta_x$. There exists a function $f_{X,j}(\cdot,\cdot)$ such that:

$$\big\| x_a(t) - x_b(t) \big\| \leq f_{X,j}(\theta_x, t - t_0) \qquad (5.149)$$

for all $x_a(t), x_b(t) \in \Omega_\rho$, and $u_i^c(t), u_i^{c-1} \in U_i$ and $\|u_i^c(t) - u_i^{c-1}(t)\| \leq \Delta u_i$ ($i = 1, \ldots, m$) with:

$$f_{X,j}(\tau) = \left(\frac{C_{2,j}}{C_{1,j}} + \theta_x \right)\big(e^{C_{1,j}\tau} - 1\big), \qquad (5.150)$$

where $C_{1,j} = L_x + \sum_{i=1}^{m,\ i\neq j} L_{g_i}u_i^{\max}$ and $C_{2,j} = \sum_{i=1}^{m,\ i\neq j} M_{g_i}\Delta u_i$.

Proof Define the error vector as $e(t) = x_a(t) - x_b(t)$. The time derivative of the error is:

$$\dot{e}(t) = f\big(x_a(t)\big) - f\big(x_b(t)\big) + \sum_{i=1}^{m,\ i\neq j} g_i\big(x_a(t)\big)u_i^c(t) - \sum_{i=1}^{m,\ i\neq j} g_i\big(x_b(t)\big)u_i^{c-1}(t).$$

$$(5.151)$$

Adding and subtracting $\sum_{i=1}^{m,\ i\neq j} g_i(x_b(t))u_i^c(t)$ to/from the right-hand side of the above equation, we obtain the following equation:

$$\dot{e}(t) = f\big(x_a(t)\big) - f\big(x_b(t)\big) + \sum_{i=1}^{m,\ i\neq j} \big(g_i\big(x_a(t)\big)u_i^c(t) - g_i\big(x_b(t)\big)u_i^c(t)\big)$$

$$+ \sum_{i=1}^{m,\ i\neq j} \big(g_i\big(x_b(t)\big)u_i^c(t) - g_i\big(x_b(t)\big)u_i^{c-1}(t)\big). \qquad (5.152)$$

From the Lipschitz properties of Eqs. 5.10–5.12, the fact that the manipulated inputs are bounded in convex sets and the difference between $u_i^c(t)$ and $u_i^{c-1}(t)$ is bounded, the following inequality can be obtained:

$$\big\|\dot{e}(t)\big\| \leq L_x\big\|x_a(t) - x_b(t)\big\| + \sum_{i=1}^{m,\ i\neq j} L_{u_i}\big\|x_a(t) - x_b(t)\big\|\,\big\|u_i^c(t)\big\|$$

$$+ \sum_{i=1}^{m,\ i\neq j} \big\|g_i\big(x_b(t)\big)\big\|\,\big\|u_i^c(t) - u_i^{c-1}(t)\big\|$$

$$\leq L_x \|e(t)\| + \sum_{i=1}^{m,\ i \neq j} L_{u_i} u_i^{\max} \|e(t)\| + \sum_{i=1}^{m,\ i \neq j} M_{g_i} \Delta u_i. \qquad (5.153)$$

Denoting $C_{1,j} = L_x + \sum_{i=1}^{m,\ i \neq j} L_{g_i} u_i^{\max}$ and $C_{2,j} = \sum_{i=1}^{m,\ i \neq j} M_{g_i} \Delta u_i$, we can obtain:

$$\|\dot{e}(t)\| \leq C_{1,j} \|e(t)\| + C_{2,j}. \qquad (5.154)$$

Integrating $\|\dot{e}(t)\|$ with initial condition $\|e(t_0)\| = \|n_x\|$ (recall that $x_b(t_0) = x_a(t_0) + n_x$) and taking into account that $\|n_x\| \leq \theta_x$, the following bound on the norm of the error vector is obtained:

$$\|e(t)\| \leq \left(\frac{C_{2,j}}{C_{1,j}} + \theta_x\right)\left(e^{C_{1,j}(t-t_0)} - 1\right). \qquad (5.155)$$

This implies that Eq. 5.149 holds for:

$$f_{X,j}(\theta_x, \tau) = \left(\frac{C_{2,j}}{C_{1,j}} + \theta_x\right)\left(e^{C_{1,j}\tau} - 1\right). \qquad (5.156)$$

$\square$

Proposition 5.1 bounds the difference between the nominal state trajectory under the optimized control inputs and the predicted nominal state trajectory generated in each LMPC optimization problem. To simplify the proof of Theorem 5.4, we define a new function $f_X(\tau)$ based on $f_{X,i}$, $i = 1, \ldots, m$, as follows:

$$f_X(\tau) = \sum_{i=1}^{m} \left(\frac{1}{m} L_x' + L_{u_i}' u_i^{\max}\right)\left(\frac{1}{C_{1,i}} f_{X,i}(0, \tau) - \frac{C_{2,i}}{C_{1,i}}\tau\right). \qquad (5.157)$$

It is easy to verify that $f_X(\tau)$ is a strictly increasing and convex function of its argument. In Theorem 5.4 below, we provide sufficient conditions under which the iterative DMPC of Eqs. 5.139–5.146 guarantees that the state of the closed-loop system is ultimately bounded in a region that contains the origin.

Theorem 5.4 *Consider the system of Eq. 5.1 in closed-loop with x available at asynchronous sampling time instants $\{t_{a \geq 0}\}$, satisfying the condition of Eq. 2.22, under the DMPC design of Eqs. 5.139–5.146 based on a control law $h(x)$ that satisfies the conditions of Eqs. 4.3–5.8. Let $\Delta, \varepsilon_s > 0$, $\rho > \rho_{\min} > 0$, $\rho > \rho_s > 0$ and $N \geq N_R \geq 1$ satisfy the conditions of Eqs. 5.108 and the following inequality:*

$$-N_R \varepsilon_s + f_X(N_R \Delta) + f_V\left(f_W(0, N_R \Delta)\right) < 0 \qquad (5.158)$$

with f_X defined in Eq. 5.157, f_V defined in Eq. 2.49, f_W defined in Eq. 5.119, and N_R being the smallest integer satisfying $N_R \Delta \geq T_m$. If the initial state of the closed-loop system $x(t_0) \in \Omega_\rho$, then $x(t)$ is ultimately bounded in $\Omega_{\rho_b} \subseteq \Omega_\rho$ where:

$$\rho_b = \rho_{\min} + f_X(N_R \Delta) + f_V\left(f_W(0, N_R \Delta)\right) \qquad (5.159)$$

with $\rho_{\min}$ defined in Eq. 5.26.

Proof We follow a similar strategy to the one in the proof of Theorem 5.3. In order to simplify the notation, we assume that all the signals used in this proof refer to the different optimization variables of the problems solved at t_a with the initial condition $x(t_a)$. This proof also includes two parts.

Part I: In this part, we prove that the stability results stated in Theorem 5.4 hold in the case that $t_{a+1} - t_a = T_m$ for all a and $T_m = N_R \Delta$. The derivative of the Lyapunov function of the nominal system of Eq. 5.1 under the control of the iterative DMPC of Eqs. 5.139–5.146 from t_a to t_{a+1} is expressed as follows:

$$\dot{V}(\tilde{x}(t)) = \frac{\partial V(\tilde{x}(t))}{\partial x}\left(f(\tilde{x}(t)) + \sum_{i=1}^{m} g_i(\tilde{x}(t))u_{p,i}^{a,*}(t|t_a) \right), \quad \forall t \in [t_a, t_a + N_R \Delta).$$

(5.160)

Adding the above equation and the constraints of Eq. 5.145 in each LMPC together, we can obtain the following inequality for $t \in [t_a, t_a + N_R \Delta)$:

$$\begin{aligned}
\dot{V}(\tilde{x}(t)) \leq\ & \frac{\partial V(\tilde{x}(t))}{\partial x}\left(f(\tilde{x}(t)) + \sum_{i=1}^{m} g_i(\tilde{x}(t))u_{p,i}^{a,*}(t|t_a) \right) \\
& + \frac{\partial V(\hat{x}(t|t_a))}{\partial x}\left(f(\hat{x}(t|t_a)) + \sum_{i=1}^{m} g_i(\hat{x}(t|t_a))u_{n,i}(t|t_a) \right) \\
& - \frac{\partial V(\tilde{x}^1(t))}{\partial x}\left(\frac{1}{m}f(\tilde{x}^1(t)) + g_1(\tilde{x}^1(t))u_{p,1}^{a,*}(t|t_k) \right) \\
& - \cdots \\
& - \frac{\partial V(\tilde{x}^m(t))}{\partial x}\left(\frac{1}{m}f(\tilde{x}^m(t)) + g_m(\tilde{x}^m(t))u_{p,m}^{a,*}(t|t_a) \right).
\end{aligned}$$

(5.161)

Reworking the above inequality, the following inequality can be obtained for $t \in [t_a, t_a + N_R \Delta)$:

$$\begin{aligned}
\dot{V}(\tilde{x}(t)) \leq\ & \frac{\partial V(\hat{x}(t|t_a))}{\partial x}\left(f(\hat{x}(t|t_a)) + \sum_{i=1}^{m} g_i(\hat{x}(t|t_a))u_{n,i}(t|t_a) \right) \\
& + \frac{\partial V(\tilde{x}(t))}{\partial x}\left(\frac{1}{m}f(\tilde{x}(t)) + g_1(\tilde{x})u_{p,1}^{a,*}(t|t_a) \right) \\
& - \frac{\partial V(\tilde{x}^1(t))}{\partial x}\left(\frac{1}{m}f(\tilde{x}^1(t)) + g_1(\tilde{x}^1(t))u_{p,1}^{a,*}(t|t_a) \right) \\
& + \cdots \\
& + \frac{\partial V(\tilde{x}(t))}{\partial x}\left(\frac{1}{m}f(\tilde{x}(t)) + g_m(\tilde{x})u_{p,m}^{a,*}(t|t_a) \right) \\
& - \frac{\partial V(\tilde{x}^m(t))}{\partial x}\left(\frac{1}{m}f(\tilde{x}^m(t)) + g_m(\tilde{x}^m(t))u_{p,m}^{a,*}(t|t_a) \right).
\end{aligned}$$

(5.162)

(5.163)

By the continuity and locally Lipschitz properties of Eqs. 5.14–5.15, the following inequality can be obtained for $t \in [t_a, t_a + N_R \Delta)$:

$$\dot{V}(\tilde{x}(t)) \leq \dot{V}(\hat{x}(t|t_a)) + \left(\frac{1}{m}L'_x + L'_{u_1}u^{a,*}_{p,1}(t|t_a)\right)\|\tilde{x}(t) - \tilde{x}^1(t)\| + \cdots$$

$$+ \left(\frac{1}{m}L'_x + L'_{u_m}u^{a,*}_{p,m}(t|t_a)\right)\|\tilde{x}(t) - \tilde{x}^m(t)\| \qquad (5.164)$$

Applying Proposition 5.1 to the above inequality of Eq. 5.164, we obtain the following inequality:

$$\dot{V}(\tilde{x}(t)) \leq \dot{V}(\hat{x}(t|t_a)) + \left(\frac{1}{m}L'_x + L'_{u_1}u^{max}_1\right)f_{X,1}(0, t - t_a) + \cdots$$

$$+ \left(\frac{1}{m}L'_x + L'_{u_m}u^{max}_m\right)f_{X,m}(0, t - t_a). \qquad (5.165)$$

Integrating the inequality of Eq. 5.165 from $t = t_a$ to $t = t_a+ = N_R\Delta$ and taking into account that $\tilde{x}(t_a) = \hat{x}(t_a)$ and $t_{a+1} - t_a = N_R\Delta$, the following inequality can be obtained:

$$V(\tilde{x}(t_{a+1})) \leq V(\hat{x}(t_{a+1}))$$

$$+ \left(\frac{1}{m}L'_x + L'_{u_1}u^{max}_1\right)\left(\frac{1}{C_{1,1}}f_{X,1}(0, N_R\Delta) - \frac{C_{2,1}}{C_{1,1}}N_R\Delta\right) + \cdots$$

$$+ \left(\frac{1}{m}L'_x + L'_{u_m}u^{max}_m\right)\left(\frac{1}{C_{1,m}}f_{X,m}(0, N_R\Delta) - \frac{C_{2,m}}{C_{1,m}}N_R\Delta\right). \qquad (5.166)$$

From the definition of $f_X(\cdot)$, we have

$$V(\tilde{x}(t_{a+1})) \leq V(\hat{x}(t_{a+1})) + f_X(N_R\Delta). \qquad (5.167)$$

By Corollaries 5.2 and 5.3 and following similar calculations to the ones in the proof of Theorem 5.3, we obtain the following inequality

$$V(x(t_{a+1})) \leq \max\{V(x(t_a)) - N_R\varepsilon_s, \rho_{min}\} + f_X(N_R\Delta) + f_V(f_W(0, N_R\Delta)). \qquad (5.168)$$

If the condition of Eq. 5.158 is satisfied, we know that there exists $\varepsilon_w > 0$ such that the following inequality holds:

$$V(x(t_{a+1})) \leq \max\{V(x(t_a)) - \varepsilon_w, \rho_b\}, \qquad (5.169)$$

which implies that if $x(t_a) \in \Omega_\rho/\Omega_{\rho_b}$, then $V(x(t_{a+1})) < V(x(t_a))$, and if $x(t_a) \in \Omega_{\rho_b}$, then $V(x(t_{a+1})) \leq \rho_b$.

Because the upper bound on the difference between the Lyapunov function of the actual trajectory x and the nominal trajectory $\tilde{x}$ is a strictly increasing function of time, the inequality of Eq. 5.169 also implies that:

$$V(x(t)) \leq \max\{V(x(t_a)) - \varepsilon_w, \rho_b\}, \quad \forall t \in [t_a, t_{a+1}]. \tag{5.170}$$

Using the inequality of Eq. 5.170 recursively, it can be proved that if $x(t_0) \in \Omega_\rho$, then the closed-loop trajectories of the system of Eq. 5.1 under the iterative DMPC design stay in Ω_ρ for all times (i.e., $x(t) \in \Omega_\rho$ for all t). Moreover, if $x(t_0) \in \Omega_\rho$, the closed-loop trajectories of the system of Eq. 5.1 under the iterative DMPC design satisfy:

$$\limsup_{t \to \infty} V(x(t)) \leq \rho_b. \tag{5.171}$$

This proves that $x(t) \in \Omega_\rho$ for all times and $x(t)$ is ultimately bounded in Ω_{ρ_b} for the case when $t_{a+1} - t_a = T_m$ for all a and $T_m = N_R \Delta$.

Part 2: In this part, we extend the results proved in Part 1 to the general case, that is, $t_{a+1} - t_a \leq T_m$ for all a and $T_m \leq N_R \Delta$ which implies that $t_{a+1} - t_a \leq N_R \Delta$. Because f_V, f_W and f_X are strictly increasing functions of time and f_X, f_V are convex, following similar steps as in Part 1, it can be shown that the inequality of Eq. 5.168 still holds. This proves that the stability results stated in Theorem 5.4 hold. $\square$

Remark 5.15 Referring to the design of the LMPC of Eqs. 5.139, the constraint of Eq. 5.142 ensures that the deviation of the predicted future state evolution (using input trajectories obtained in the last iteration) from the actual system state evolution is bounded. It also ensures that the results stated in Theorem 5.4 do not depend on the iteration number c which means the iterations of the DMPC can be terminated at any iteration and the stability properties stated in Theorem 5.4 continue to hold. The constraint of Eq. 5.142 can be also imposed as the termination condition of the iterative DMPC; that is, the DMPC stops iterating when $\|u_{p,i}(t) - u_{p,i}^{a,c-1}(t|t_a)\| \leq \Delta u_i$, $i = 1, \ldots, m$, for all $t \in [t_a, t_a + N_R \Delta)$. In this case, however, the stability properties stated in Theorem 5.4 have dependence on the iteration number c in a way that they hold only after the termination condition of Eq. 5.142 is satisfied.

5.5.4 Application to an Alkylation of Benzene Process

Consider the alkylation of benzene with ethylene process of Eqs. 5.56–5.80 described in Sect. 5.4.3. The control objective is still to drive the system from the initial condition as shown in Table 5.6 to the desired steady-state as shown in Table 5.4. The manipulated inputs are the heat injected to or removed from the five vessels, Q_1, Q_2, Q_3, Q_4 and Q_5, and the feed stream flow rates to CSTR-2 and CSTR-3, F_4 and F_6, whose steady-state input values are shown in Table 5.3. We design three distributed LMPCs to manipulate the 7 inputs. Similarly, the first distributed controller (LMPC 1) will be designed to decide the values of Q_1, Q_2 and

Q_3, the second distributed controller (LMPC 2) will be designed to decide the values of Q_4 and Q_5, and the third distributed controller (LMPC 3) will be designed to decide the values of F_4 and F_6. The deviations of these inputs from their corresponding steady-state values are subject to the constraints shown in Table 5.5. We use the same design of $h(x)$ as in Sect. 5.4.3 with a quadratic Lyapunov function $V(x) = x^T P x$ with P being the following weight matrix:

$$P = diag\big([1\ 1\ 1\ 1\ 1\ 1\ 1\ 1\ 1\ 1\ 1\ 1\ 1\ 1\ 1\ 1\ 1\ 1\ 1\ 1\ 1\ 1\ 1\ 1\ 1]\big). \qquad (5.172)$$

Based on $h(x)$, we design the sequential DMPC of Eqs. 5.98–5.106 and the iterative DMPC of Eqs. 5.139–5.146 with the following weighting matrices:

$$Q_c = diag\big([1\ 1\ 1\ 1\ 10^3\ 1\ 1\ 1\ 1\ 10^3\ 10\ 10\ 10\ 10\ 3000\ 1\ 1\ 1\ 1\ 10^3\ 1\ 1\ 1\ 1\ 10^3]\big),$$
$$(5.173)$$

and $R_{c1} = diag([1 \times 10^{-8}\ 1 \times 10^{-8}\ 1 \times 10^{-8}])$, $R_{c2} = diag([1 \times 10^{-8}\ 1 \times 10^{-8}])$ and $R_{c3} = diag([10\ 10])$. The sampling time of the LMPCs is chosen to be $\Delta = 30$ s. For the iterative DMPC of Eqs. 5.139–5.146, Δu_i is chosen to be $0.25 u_i^{\max}$ for all the distributed LMPCs and maximum iteration numbers (i.e., $c \leq c_{\max}$) are applied as the termination conditions. In all the simulations, bounded process noise is added to the right hand side of the ordinary differential equations of the process model to simulate disturbances/model uncertainty.

We consider that the state of the process of Eqs. 5.56–5.80 is sampled asynchronously and that the maximum interval between two consecutive measurements is $T_m = 75$ s. The asynchronous nature of the measurements is introduced by the measurement difficulties of the full state given the presence of several species concentration measurements. We will compare the sequential and iterative DMPC for systems subject to asynchronous measurements with a centralized LMPC which takes into account asynchronous measurements explicitly as presented in Sect. 2.7. The centralized LMPC uses the same weighting matrices, sampling time and prediction horizon as used in the DMPCs. To model the time sequence $\{t_{a \geq 0}\}$, we apply an upper bounded random Poisson process. The Poisson process is defined by the number of events per unit time W. The interval between two successive state sampling times is given by $\Delta_a = \min\{-\ln \chi / W, T_m\}$, where χ is a random variable with uniform probability distribution between 0 and 1. This generation ensures that $\max_a \{t_{a+1} - t_a\} \leq T_m$. In the simulations, W is chosen to be 30 and the time sequence generated by this bounded Poisson process is shown in Fig. 5.8. For this set of simulations, we choose the prediction horizon of all the LMPCs to be $N = 3$ and choose $N_R = N$ so that $N_R \Delta \geq T_m$.

We first compare the DMPC designs for systems subject to asynchronous measurements with the centralized LMPC from a stability point of view. Figure 5.9 shows the trajectory of the Lyapunov function $V(x)$ under these control designs. From Fig. 5.9, we see that the DMPC designs as well as the centralized LMPC design are able to drive the system state to a region very close to the desired steady state. From Fig. 5.9, we can also see that the sequential DMPC, the centralized LMPC and the iterative DMPC with $c_{\max} = 5$ give very similar trajectories of $V(x)$. Another important aspect we can see from Fig. 5.9(b) is that at the early stage of the

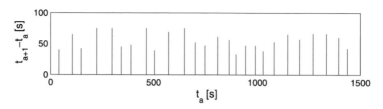

Fig. 5.8 Asynchronous measurement sampling times $\{t_{a \geq 0}\}$ with $T_m = 75$ s: the x-axis indicates $\{t_{a \geq 0}\}$ and the y-axis indicates the size of the interval between t_a and t_{a-1}

closed-loop system simulation, because of the strong driving force related to the difference between the set-point and the initial condition, the process noise/disturbance has small influence on the process dynamics, even though the controller(s) has/have to operate in the presence of asynchronous measurements. When the states are getting close to the set-point, the Lyapunov function starts to fluctuate due to the domination of noise/disturbance over the vanishing driving force. However, the DMPC designs are able to maintain practical stability of the closed-loop system and keep the trajectory of the Lyapunov function in a bounded region ($V(x) \leq 250$) very close to the steady state.

Next, we compare the evaluation times of the LMPCs in these control designs. The simulations are carried out by JAVA™ programming language in a PENTIUM® 3.20 GHz computer. The optimization problems are solved by the open source interior point optimizer Ipopt [109]. We evaluate the LMPC optimization problems for 100 runs. The mean evaluation time of the centralized LMPC is about 23.7 s. The mean evaluation time for the sequential DMPC scheme, which is the sum of the evaluation times (1.9 s, 3.6 s and 3.2 s) of the three LMPCs, is about 8.7 s. The mean evaluation time of the iterative DMPC scheme with one iteration is 6.3 s which is the largest evaluation time among the evaluation times (1.6 s, 6.3 s and 4.3 s) of the three LMPCs. The mean evaluation time of the iterative DMPC architecture with four iterations is 18.7 s with the evaluation times of the three LMPCs being 6.9 s, 18.7 s and 14.0 s. From this set of simulations, we see that the DMPC designs lead to a significant reduction in the controller evaluation time compared with a centralized LMPC design though they provide a very similar performance.

5.6 Iterative DMPC Design with Delayed Measurements

In this section, we consider the design of DMPC for systems subject to delayed measurements. In Chap. 4, we pointed out that in order to obtain a good estimate of the current system state from a delayed state measurement, a DMPC design should have bi-directional communication among the distributed controllers. Consequently, we focus on the design of DMPC for nonlinear systems subject to delayed measurements in an iterative DMPC framework.

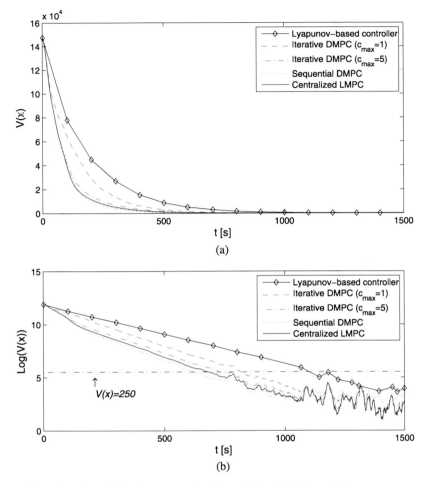

Fig. 5.9 Trajectories of the Lyapunov function of the alkylation of benzene process of Eqs. 5.56–5.80 under the nonlinear control law $h(x)$ implemented in a sample-and-hold fashion and with open-loop state estimation, the iterative DMPC of Eqs. 5.139–5.146 with $c_{max} = 1$ and $c_{max} = 5$, the sequential DMPC of Eqs. 5.98–5.106 and the centralized LMPC accounting for asynchronous measurements: (**a**) $V(x)$; (**b**) $\text{Log}(V(x))$

5.6.1 Modeling of Delayed Measurements

We assume that the state of the system of Eq. 5.1 is received by the controllers at asynchronous time instants t_a where $\{t_{a \geq 0}\}$ is a random increasing sequence of times and that there exists an upper bound T_m on the interval between two successive measurements. We also assume that there are delays in the measurements received by the controllers due to delays in the sampling process and data transmission. In order to model delays in measurements, another auxiliary variable d_a is introduced to indicate the delay corresponding to the measurement received at time t_a, that is,

Fig. 5.10 Iterative DMPC
for nonlinear systems subject
to delayed measurements

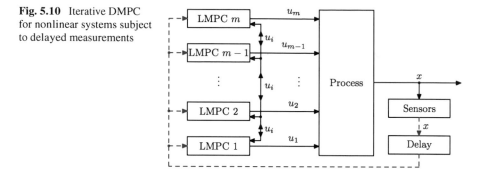

at time t_a, the measurement $x(t_a - d_a)$ is received. In order to study the stability
properties in a deterministic framework, we assume that the delays associated with
the measurements are smaller than an upper bound D.

5.6.2 Iterative DMPC Formulation

As in the DMPC designs for systems subject to asynchronous measurements, we
take advantage of the system model both to estimate the current system state from a
delayed measurement and to control the system in open-loop when new information
is not available. To this end, when a delayed measurement is received, the distributed
controllers use the system model and the input trajectories that have been applied
to the system to get an estimate of the current state and then based on the estimate,
MPC optimization problems are solved to compute the optimal future input trajectory that will be applied until new measurements are received. A schematic of the
iterative DMPC for systems subject to delayed measurements is shown in Fig. 5.10.
The implementation strategy for the iterative DMPC design is as follows:

1. When a measurement $x(t_a - d_a)$ is available at t_a, all the distributed controllers
   receive the state measurement and check whether the measurement provides new
   information. If $t_a - d_a > \max_{l<a} t_l - d_l$, go to Step 2. Else the measurement does
   not contain new information and is discarded, go to Step 3.
2. All the distributed controllers estimate the current state of the system $x^e(t_a)$ and
   then evaluate their future input trajectories in an iterative fashion with initial
   input guesses generated by $h(\cdot)$.
3. At iteration c ($c \geq 1$):
   3.1. Each controller evaluates its own future input trajectory based on $x^e(t_a)$ and
        the latest received input trajectories of all the other distributed controllers
        (when $c = 1$, initial input guesses generated by $h(\cdot)$ are used).
   3.2. The controllers exchange their future input trajectories. Based on all the
        input trajectories, each controller calculates and stores the value of the cost
        function.

4. If a termination condition is satisfied, each controller sends its entire future input trajectory corresponding to the smallest value of the cost function to its actuators; if the termination condition is not satisfied, go to Step 3 ($c \leftarrow c + 1$).
5. When a new measurement is received ($a \leftarrow a + 1$), go to Step 1.

In order to estimate the current system state $x^e(t_a)$ based on a delayed measurement $x(t_a - d_a)$, the distributed controllers take advantage of the input trajectories that have been applied to the system from $t_a - d_a$ to t_a and the system model of Eq. 5.1. Let us denote the input trajectories that have been applied to the system as $u^*_{d,i}(t)$, $i = 1, \ldots, m$. Therefore, $x^e(t_a)$ is evaluated by integrating the following equation:

$$\dot{x}^e(t) = f\big(x^e(t)\big) + \sum_{i=1}^{m} g_i\big(x^e(t)\big) u^*_{d,i}(t), \quad \forall t \in [t_a - d_a, t_a) \qquad (5.174)$$

with $x^e(t_a - d_a) = x(t_a - d_a)$.

Before going to the design of the iterative DMPC, we need to define another nominal sampled trajectory $\check{x}(t|t_a)$ for $t \in [t_a, t_a + N\Delta)$, which is obtained by replacing $\hat{x}(t|t_a)$ with $\check{x}(t|t_a)$ in Eq. 5.36 and then integrating the equation with $\check{x}(t_a|t_a) = x^e(t_a)$. Based on $\check{x}(t|t_a)$, we define a new input trajectory as follows:

$$u^e_{n,j}(t|t_a) = h_j\big(\check{x}(t_a + l\Delta|t_a)\big),$$

$$j = 1, \ldots, m, \ \forall t \in \big[t_a + l\Delta, t_a + (l+1)\Delta\big), l = 0, \ldots, N - 1, \qquad (5.175)$$

which will be used in the design of the LMPC to construct the stability constraint and used as the initial input guess for iteration 1 (i.e., $u^{*,0}_{d,i} = u^e_{n,i}$ for $i = 1, \ldots, m$).

Specifically, the design of LMPC j, $j = 1, \ldots, m$, at iteration c is based on the following optimization problem:

$$\min_{u_j \in S(\Delta)} \int_{t_a}^{t_a + N\Delta} \left[\left\| \tilde{x}^j(\tau) \right\|_{Q_c} + \sum_{i=1}^{m} \left\| u_i(\tau) \right\|_{R_{ci}} \right] d\tau, \qquad (5.176)$$

$$\text{s.t.} \quad \dot{\tilde{x}}^j(t) = f\big(\tilde{x}^j(t)\big) + \sum_{i=1}^{m} g_i\big(\tilde{x}^j(t)\big) u_i(t), \qquad (5.177)$$

$$u_i(t) = u^{*,c-1}_{d,i}(t|t_a), \quad \forall i \neq j, \qquad (5.178)$$

$$\left\| u_j(t) - u^{*,c-1}_{d,j}(t|t_a) \right\| \leq \Delta u_j, \quad \forall \tau \in [t_a, t_a + N_{D,a}\Delta), \qquad (5.179)$$

$$u_j(\tau) \in U_j, \qquad (5.180)$$

$$\tilde{x}^j(t_a) = x^e(t_a), \qquad (5.181)$$

$$\frac{\partial V(\tilde{x}^j(t))}{\partial \tilde{x}^j} \left(\frac{1}{m} f\big(\tilde{x}^j(t)\big) + g_j\big(\tilde{x}^j(t)\big) u_j(t) \right)$$

$$\leq \frac{\partial V(\check{x}(t|t_a))}{\partial \check{x}} \left(\frac{1}{m} f\big(\check{x}(t|t_a)\big) + g_j\big(\check{x}(t|t_a)\big) u^e_{n,j}(t|t_a) \right),$$

$$\forall \tau \in [t_a, t_a + N_{D,a}\Delta), \qquad (5.182)$$

where $N_{D,a}$ is the smallest integer satisfying $N_{D,a}\Delta \geq T_m + D - d_a$. The optimal solution to this optimization problem is denoted $u_{d,j}^{*,c}(a|t_a)$ which is defined for $t \in [t_a, t_a + N\Delta)$. Accordingly, we define the final optimal input trajectory of LMPC j of Eqs. 5.176–5.182 as $u_{d,j}^*(t|t_k)$ which is also defined for $t \in [t_a, t_a + N\Delta)$. Note again that the length of the constraint $N_{D,a}$ depends on the current delay d_a, so it may have different values at different time instants and has to be updated before solving the optimization problems.

The manipulated inputs of the closed-loop system under the above iterative DMPC for systems subject to delayed measurements are defined as follows:

$$u_i(t) = u_{d,i}^*(t|t_a), \quad i = 1, \ldots, m, \forall t \in [t_a, t_{a+q}) \qquad (5.183)$$

for all t_a such that $t_a - d_a > \max_{l<a} t_l - d_l$ and for a given t_a, the variable q denotes the smallest integer that satisfies $t_{a+q} - d_{a+q} > t_a - d_a$.

5.6.3 Stability Properties

The stability properties of the iterative DMPC of Eqs. 5.176–5.183 are stated in the following theorem.

Theorem 5.5 *Consider the system of Eq. 5.1 in closed-loop with x available at asynchronous sampling time instants $\{t_{a\geq0}\}$ involving time-varying delays such that $d_a \leq D$ for all $a \geq 0$, satisfying the condition of Eq. 2.22, under the iterative DMPC of Eqs. 5.176–5.183 based on a control law $u = h(x)$ that satisfies the conditions of Eqs. 5.5–5.8. Let $\Delta, \varepsilon_s > 0, \rho > \rho_{\min} > 0, \rho > \rho_s > 0, N \geq 1$ and $D \geq 0$ satisfy the condition of Eq. 5.108 and the following inequality:*

$$-N_R\varepsilon_s + f_X(N_D\Delta) + f_V\big(f_W(0, N_D\Delta)\big) + f_V\big(f_W(0, D)\big) < 0 \qquad (5.184)$$

with f_V defined in Eq. 2.49, f_W defined in Eq. 5.119, N_D being the smallest integer satisfying $N_D\Delta \geq T_m + D$ and N_R being the smallest integer satisfying $N_R\Delta \geq T_m$. If the initial state of the closed-loop system $x(t_0) \in \Omega_\rho$, $N \geq N_D$ and $d_0 = 0$, then $x(t)$ is ultimately bounded in $\Omega_{\rho_d} \subseteq \Omega_\rho$ where:

$$\rho_d = \rho_{\min} + f_X(N_D\Delta) + f_V\big(f_W(0, N_D\Delta)\big) + f_V\big(f_W(0, D)\big) \qquad (5.185)$$

with $\rho_{\min}$ defined in Eq. 5.26.

Proof We assume that at t_a, a delayed measurement $x(t_a - d_a)$ containing new information is received, and that the next measurement with new state information is not received until t_{a+i}. This implies that $t_{a+i} - d_{a+i} > t_a - d_a$ and that the iterative DMPC of Eqs. 5.176–5.183 is solved at t_a and the optimal input trajectories

$u_{d,i}^*(t|t_a)$, $i = 1, \ldots, m$, are applied from t_a to t_{a+i}. In this proof, we will refer to $\tilde{x}(t)$ for $t \in [t_a, t_{a+i})$ as the state trajectory of the nominal system of Eq. 5.1 under the control of the iterative DMPC of Eqs. 5.176–5.183 with $\tilde{x}(t_a) = x^e(t_a)$.

Part I: In this part, we prove that the stability results stated in Theorem 5.5 hold for $t_{a+i} - t_a = N_{D,a}\Delta$ and all $d_a \leq D$. By Corollary 5.2 and taking into account that $\check{x}(t_a) = x^e(t_a)$, the following inequality can be obtained:

$$V(\check{x}(t_{a+i})) \leq \max\{V(x^e(t_a)) - N_{D,a}\varepsilon_s, \rho_{\min}\}. \tag{5.186}$$

By Corollary 5.3 and taking into account that $x^e(t_a - d_a) = x(t_a - d_a)$, $\tilde{x}(t_a) = x^e(t_a)$ and $N_D\Delta \geq N_{D,a}\Delta + d_a$, the following inequalities can be obtained:

$$\|x^e(t_a) - x(t_a)\| \leq f_W(0, d_a), \tag{5.187}$$

$$\|\tilde{x}(t_{a+i}) - x(t_{a+i})\| \leq f_W(0, N_D\Delta). \tag{5.188}$$

When $x(t) \in \Omega_\rho$ for all times (this point will be proved below), we can apply Proposition 2.3 to obtain the following inequalities:

$$V(x^e(t_a)) \leq V(x(t_a)) + f_V(f_W(0, d_a)), \; V(x(t_{a+i}))$$
$$\leq V(\tilde{x}(t_{a+i})) + f_V(f_W(0, N_D\Delta)). \tag{5.189}$$

From Eqs. 5.186 and 5.189, the following inequality is obtained:

$$V(\check{x}(t_{a+i})) \leq \max\{V(x(t_a)) - N_{D,a}\varepsilon_s, \rho_{\min}\} + f_V(f_W(0, d_a)). \tag{5.190}$$

By Proposition 5.1 and following similar steps as in the proof of Theorem 5.4, the following inequality can be obtained:

$$V(\tilde{x}(t_{a+i})) \leq V(\check{x}(t_{a+i})) + f_X(N_{D,a}\Delta). \tag{5.191}$$

From Eqs. 5.189, 5.190 and 5.191, the following inequality is obtained:

$$V(x(t_{a+i})) \leq \max\{V(x(t_a)) - N_{D,a}\varepsilon_s, \rho_{\min}\} + f_V(f_W(0, d_a))$$
$$+ f_V(f_W(0, N_D\Delta)) + f_X(N_{D,a}\Delta). \tag{5.192}$$

In order to prove that the Lyapunov function is decreasing between two consecutive new measurements, the following inequality must hold:

$$N_{D,a}\varepsilon_s > f_V(f_W(0, d_a)) + f_V(f_W(0, N_D\Delta)) + f_X(N_{D,a}\Delta) \tag{5.193}$$

for all possible $0 \leq d_a \leq D$. Taking into account that f_W, f_V and f_X are strictly increasing functions of time, $N_{D,a}$ is a decreasing function of the delay d_a and that if $d_a = D$ then $N_{D,a} = N_R$, then if the condition of Eq. 5.184 is satisfied, the condition of Eq. 5.193 holds for all possible d_a and there exists $\varepsilon_w > 0$ such that the following inequality holds:

$$V(x(t_{a+i})) \leq \max\{V(x(t_a)) - \varepsilon_w, \rho_d\}, \tag{5.194}$$

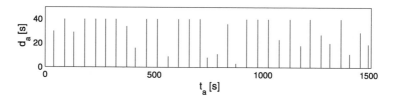

Fig. 5.11 Asynchronous time sequence $\{t_{a\geq0}\}$ and corresponding delay sequence $\{d_{a\geq0}\}$ with $T_m = 50$ s and $D = 40$ s: the x-axis indicates $\{t_{a\geq0}\}$ and the y-axis indicates the size of d_a

which implies that if $x(t_a) \in \Omega_\rho/\Omega_{\rho_d}$, then $V(x(t_{a+i})) < V(x(t_a))$, and if $x(t_a) \in \Omega_{\rho_d}$, then $V(x(t_{a+i})) \leq \rho_d$.

Because the upper bound on the difference between the Lyapunov function of the actual trajectory x and the nominal trajectory $\tilde{x}$ is a strictly increasing function of time, the inequality of Eq. 5.194 also implies that:

$$V\big(x(t)\big) \leq \max\big\{V\big(x(t_a)\big), \rho_d\big\}, \quad \forall t \in [t_a, t_{a+i}). \tag{5.195}$$

Using the inequality of Eq. 5.195 recursively, it can be proved that if $x(t_0) \in \Omega_\rho$, then the closed-loop trajectories of the system of Eq. 5.1 under the iterative DMPC of Eqs. 5.176–5.183 stay in Ω_ρ for all times (i.e., $x(t) \in \Omega_\rho, \forall t$). Moreover, using the inequality of Eq. 5.195 recursively, it can be proved that if $x(t_0) \in \Omega_\rho$, the closed-loop trajectories of the system of Eq. 5.1 under the iterative DMPC of Eqs. 5.176–5.183 satisfy:

$$\limsup_{t\to\infty} V\big(x(t)\big) \leq \rho_d. \tag{5.196}$$

This proves that $x(t) \in \Omega_\rho$ for all times and $x(t)$ is ultimately bounded in Ω_{ρ_d} when $t_{a+i} - t_a = N_{D,a}\Delta$.

Part 2: In this part, we extend the results proved in Part 1 to the general case, that is, $t_{a+i} - t_a \leq N_{D,a}\Delta$. Taking into account that f_V, f_W and f_X are strictly increasing functions of time and following similar steps as in Part 1, it can be readily proved that the inequality of Eq. 5.193 holds for all possible $d_a \leq D$ and $t_{a+i} - t_a \leq N_{D,a}\Delta$. Using this inequality and following the same line of argument as in the previous part, the stability results stated in Theorem 5.5 can be proved. $\qquad\square$

5.6.4 Application to an Alkylation of Benzene Process

Consider the alkylation of benzene with ethylene process of Eqs. 5.56–5.80 described in Sect. 5.4.3. We set up the simulations as described in Sect. 5.5.4.

We consider that the state of the process of Eqs. 5.56–5.80 is sampled at asynchronous time instants $\{t_{a\geq0}\}$ with an upper bound $T_m = 50$ s on the interval between two successive measurements. Moreover, we consider that there are delays involved in the measurement samplings and the upper bound on the maximum delay is $D = 40$ s. The delays in measurements can naturally arise in the context of species concentration measurements. We will compare the iterative DMPC of Eqs. 5.176–5.183 with a centralized LMPC which takes into account delayed measurements explicitly as presented in Sect. 2.8. The centralized LMPC uses the same weighting matrices, sampling time and prediction horizon as used in the DMPC. In order to model the sampling time instants, the same Poisson process as used in Sect. 5.5.4 is used to generate $\{t_{a \geq 0}\}$ with $W = 30$ and $T_m = 50$ s and another random process is used to generate the associated delay sequence $\{d_{a \geq 0}\}$ with $D = 40$ s. For this set of simulations, we also choose the prediction horizon of all the LMPCs to be $N = 3$ so that the horizon covers the maximum possible open-loop operation interval. Figure 5.11 shows the time instants when new state measurements are received and the associated delay sizes. Note that for all the control designs considered in this subsection, the same state estimation strategy shown in Eq. 5.174 is used.

Figure 5.12 shows the trajectory of the Lyapunov function $V(x)$ under different control designs. From Fig. 5.12, we see that both the iterative DMPC for systems subject to delayed measurements and the centralized LMPC accounting for delays are able to drive the system state to a region very close to the desired steady state ($V(x) \leq 250$); the trajectories of $V(x)$ generated by the iterative DMPC design are bounded by the corresponding trajectory of $V(x)$ under the nonlinear control law $h(x)$ implemented in a sample-and-hold fashion and with open-loop state estimation. From Fig. 5.12, we can also see that the centralized LMPC and the iterative DMPC with $c_{max} = 5$ give very similar trajectories of $V(x)$.

In the final set of simulations, we compare the centralized LMPC and the iterative DMPC from a performance index point of view. To carry out this comparison, the same initial condition and parameters were used for the different control schemes and the total cost under each control scheme was computed as follows:

$$J = \int_0^{t_f} \left[\|x(\tau)\|_{Q_c} + \|u_1(\tau)\|_{R_{c1}} + \|u_2(\tau)\|_{R_{c2}} + \|u_3(\tau)\|_{R_{c3}} \right] d\tau, \quad (5.197)$$

where $t_f = 1500$ s is the final simulation time. Figure 5.13 shows the total cost along the closed-loop system trajectories under the iterative DMPC of Eqs. 5.176–5.183 and the centralized LMPC accounting for delays. For the iterative DMPC design, different maximum numbers of iterations, c_{max}, are used. From Fig. 5.13, we can see that as the iteration number c increases, the performance cost given by the iterative DMPC design decreases and converges to a value which is very close to the cost of the one corresponding to the centralized LMPC. However, we note that there is no guaranteed convergence of the performance of iterative DMPC design to the performance of a centralized MPC because of the nonconvexity of the LMPC optimization problems, and the different stability constraints imposed in the centralized LMPC and the iterative DMPC design.

5.7 Handling Communication Disruptions in DMPC

In this section, we focus on a hierarchical type DMPC (see Remark 4.12) for the system of Eq. 5.1 and discuss how to handle communication disruptions—communication channel noise and data losses—between the distributed controllers.

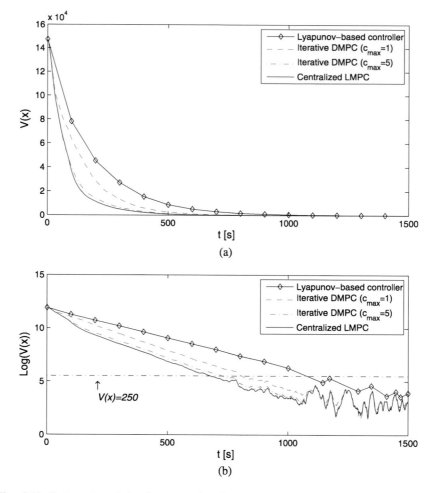

Fig. 5.12 Trajectories of the Lyapunov function of the alkylation of benzene process of Eqs. 5.56–5.80 under the nonlinear control law $h(x)$ implemented in a sample-and-hold fashion and with open-loop state estimation, the iterative DMPC of Eqs. 5.176–5.183 with $c_{max} = 1$ and $c_{max} = 5$ and the centralized LMPC accounting for delays: (**a**) $V(x)$; (**b**) $\text{Log}(V(x))$

In the sequel, we design m LMPCs to calculate the m sets of control inputs, respectively, and refer to the controller that calculates u_i $(i = 1, \ldots, m)$ as LMPC i. In this approach, LMPC 1 communicates with the rest of LMPCs (i.e., LMPC 2 to LMPC m) using one-directional communication and cooperates with them to maintain the closed-loop stability.

In the proposed design, to handle communication channel noise between the distributed controllers, feasibility problems are incorporated in the DMPC architecture to determine if the data transmitted through the communication channel is reliable or not. Based on the results of the feasibility problems, the transmitted information is accepted or rejected by LMPC 1. When there are communication data losses be-

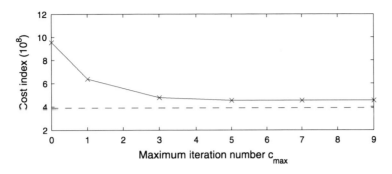

Fig. 5.13 Total performance costs along the closed-loop trajectories of the alkylation of benzene process of Eqs. 5.56–5.80 under the centralized LMPC accounting for delays (*dashed line*) and iterative DMPC of Eqs. 5.176–5.183 (*solid line*)

Fig. 5.14 Hierarchical type distributed LMPC control architecture (F means solving a feasibility problem)

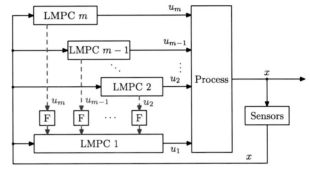

tween the distributed controllers, the closed-loop system under the proposed DMPC is guaranteed to be practically stable because of the stability constraints incorporated in the LMPC designs. A schematic diagram of the DMPC design for systems subject to communication disruptions between distributed controllers is depicted in Fig. 5.14.

5.7.1   Model of the Communication Channel

We consider data losses and channel noise in communication between the m distributed controllers. For a given input $r \in R^m$ to the communication channel, the output $\tilde{r} \in R^m$ is characterized as:

$$\tilde{r} = lr + n, \tag{5.198}$$

where l is a Bernoulli random variable with parameter α and $n \in R^m$ is a vector whose elements are white gaussian noise with zero mean and the same variance σ^2. The random variable l is used to model data losses in the communication channel. The white noise, n, is used to model channel noise, quantization error or any other

Fig. 5.15 Bounded
communication channel noise

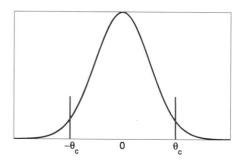

error to the transmitted signal, and it is independent of the data losses in a proba-
bilistic sense. If the receiver determines that a successful transmission is made, then
$l = 1$, otherwise $l = 0$. Furthermore, in order to get deterministic stability results,
we assume that, when a successful transmission is made, the noise, n, attached to
the input signal, r, is bounded by θ_c (that is $\|n\| \le \theta_c$) as shown in Fig. 5.15. Both
assumptions are meaningful from a practical standpoint. We further assume that the
capacity of the communication channel [15] is high enough so that we can transmit
data through it with a high rate.

Remark 5.16 Note that there are a variety of approaches to detect whether data
loss has happened at the receiver side of a communication channel. One common
approach is to measure the power of the received signal and compare it with a preconfigured signal transmission power level. If the power of the received signal is
much smaller than the preconfigured signal transmission power level, then data loss
is declared; and if the power of the received signal is close to the preconfigured
signal transmission power level, then the transmission is assumed to be successful.

5.7.2 DMPC with Communication Disruptions

The implementation strategy for the DMPC with feasibility problems is as follows:

1. At t_k, all controllers receive the sensor measurements $x(t_k)$.
2. For $i = 2, \ldots, m$
   2.1. LMPC i evaluates the optimal input trajectory of u_i based on $x(t_k)$ and
        sends the first step input values of u_i to its corresponding actuators.
   2.2. LMPC i sends the entire optimal input trajectory of u_i to LMPC 1 through
        a communication channel.
3. LMPC 1 solves a feasibility problem for each input trajectory it received to determine if the trajectory should be accepted or rejected.
4. LMPC 1 evaluates the future input trajectory of u_1 based on $x(t_k)$ and the results
   of the feasibility problems for the trajectories it received from LMPC i with
   $i = 2, \ldots, m$.
5. LMPC 1 sends the first step input value of u_1 to its corresponding actuators.

6. When a new measurement is received ($k \leftarrow k + 1$), go to Step 1.

In the sequel, we describe the design of LMPC j ($j = 2, \ldots, m$) and its corresponding feasibility problem and the design of LMPC 1. In the formulations, the input trajectories $u_{n,i}(t|t_k)$ defined in Eq. 5.37 based on the sampled state trajectory defined in Eq. 5.36 will be used.

Upon receiving the sensor measurement $x(t_k)$, LMPC j obtains its optimal input trajectory by solving the following optimization problem:

$$\min_{u_j \in S(\Delta)} \int_{t_k}^{t_{k+N}} \left[\left\| \tilde{x}^j(\tau) \right\|_{Q_c} + \sum_{i=1}^{m} \left\| u_i(\tau) \right\|_{R_{ci}} \right] d\tau, \tag{5.199}$$

$$\dot{\tilde{x}}^j(t) = f\left(\tilde{x}^j(t)\right) + \sum_{i=1}^{m} g_i\left(\tilde{x}^j(t)\right) u_i(t), \tag{5.200}$$

$$u_i(t) = u_{n,i}(t|t_k), \quad i \neq j, \tag{5.201}$$

$$\tilde{x}^j(t_k) = x(t_k), \tag{5.202}$$

$$u_j(t) \in U_j, \tag{5.203}$$

$$\frac{\partial V(x(t_k))}{\partial x} g_j\left(x(t_k)\right) u_j(t_k) \leq \frac{\partial V(x(t_k))}{\partial x} g_j\left(x(t_k)\right) h_j\left(x(t_k)\right), \tag{5.204}$$

where $q = 0, \ldots, N - 1$, $\tilde{x}^j$ is the predicted trajectory of the nominal system with u_j being the input trajectory computed by this LMPC j and u_i ($i \neq j$) determined by $u_{n,i}(t|t_k)$.

Let $u_j^*(t|t_k)$ denote the optimal solution of the optimization problem of Eqs. 5.199–5.204. LMPC j sends the first step value of $u_j^*(t|t_k)$ to its actuators and transmits the whole optimal input trajectory through the communication channel to LMPC 1. LMPC 1 receives a corrupted version of $u_j^*(t|t_k)$ which can be formulated as:

$$\tilde{u}_j(t|t_k) = l u_j^*(t|t_k) + n. \tag{5.205}$$

If data losses occur during the transmission of the control input trajectory from LMPC j to LMPC 1, LMPC 1 assumes that LMPC j applies $h_j(x)$ (i.e., $u_j = h_j(x)$). Note that we do not consider explicitly the step of determining whether data losses occur or not in the transmission of input trajectories. Please see Remark 5.16 on approaches of determining transmission data losses.

When a transmission of the input trajectory $u_j^*(t|t_k)$ is successful, LMPC 1 receives $\tilde{u}_j(t|t_k)$ which is a noise-corrupted version of $u_j^*(t|t_k)$. To determine the reliability of the received information, LMPC 1 solves a feasibility problem. Based on the result of the feasibility problem, LMPC 1 determines if the received information should be accepted or rejected. The feasibility problem for the information received from LMPC j is as follows:

$$\text{find} \quad z \in S(\Delta), \tag{5.206}$$

$$\tilde{u}_j(t|t_k) - \theta_c \leq z(t) \leq \tilde{u}_j(t|t_k) + \theta_c, \quad t \in [t_k, t_{k+N}), \quad (5.207)$$

$$z(t) \in U_j, \quad t \in [t_k, t_{k+N}), \quad (5.208)$$

$$\frac{\partial V(x(t_k))}{\partial x} g_j\big(x(t_k)\big) z(t_k) > g_j\big(x(t_k)\big) h_j\big(x(t_k)\big). \quad (5.209)$$

According to the bounded noise value and the received signal from the communication channel, LMPC 1 considers all the possibilities of noise effect on the optimal trajectory of LMPC j (i.e., the constraint of Eq. 5.207) and checks whether in these cases the input received from LMPC j still satisfies the stability constraint of Eq. 5.204 (i.e., the constraint of Eq. 5.209). Note that when the optimization problem of Eqs. 5.206–5.209 is not feasible, it is guaranteed that the original signal $u_j^*(t|t_k)$ after transmission through the channel still satisfies the stability constraint of Eq. 5.204. The feasibility of this problem is used to test whether there exists any possible value of the noise that could (due to corruption) end up making the implemented control action cause an increase in the Lyapunov function derivative, i.e., that $\frac{\partial V(x(t_k))}{\partial x} g_j(x(t_k))u_j(0) > g_j(x(t_k))h_j(x(t_k))$. If the problem is infeasible, it is guaranteed that the noise cannot make the control action destabilizing, and hence, the control action is accepted. On the other hand, if the problem is feasible, it opens up the possibility of the noise rendering the control action destabilizing, and hence, it is discarded. We also note that there is no requirement that θ_c is sufficient small, however, larger values of θ_c increase the range of $z(t)$ and influence the feasibility of the problem of Eqs. 5.206–5.209.

If the optimization problem of Eqs. 5.206–5.209 is not feasible, then the trajectory information received by LMPC 1 (i.e., $\tilde{u}_j(t|t_k)$) is used in the evaluation of LMPC 1; and if the optimization problem of Eqs. 5.206–5.209 is feasible, then $\tilde{u}_j(t|t_k)$ is discarded and the input trajectory $u_{n,j}(t|t_k)$ will be used in the evaluation of LMPC 1. If we define the trajectory of u_j that is used in the evaluation of LMPC 1 as $\tilde{u}_j^*(t|t_k)$, then it is defined as follows:

$$\tilde{u}_j^*(t|t_k) = \begin{cases} \tilde{u}_j(t|t_k) & \text{if the problem of Eqs. 5.206–5.209 is not feasible} \\ & \text{and there is no data loss,} \\ u_{n,j}(t|t_k) & \text{if the problem of Eqs. 5.206–5.209 is feasible or} \\ & \text{there exists data loss.} \end{cases}$$

Note that when data loss in the communication channel occurs, the input trajectory $u_{n,j}(t|t_k)$ is also used in the evaluation of LMPC 1. Note also that the above strategy on the use of the corrupted communication information is just one of many possible options to handle communication disruptions in the DMPC architecture.

Employing $\tilde{u}_j^*$, $j = 2, \ldots, m$, LMPC 1 obtains its optimal trajectory according to the following optimization problem:

$$\min_{u_1 \in S(\Delta)} \int_{t_k}^{t_{k+N}} \left[\left\| \tilde{x}^1(\tau) \right\|_{Q_c} + \sum_{i=1}^{m} \left\| u_i(\tau) \right\|_{R_{ci}} \right], \quad (5.210)$$

$$\dot{\tilde{x}}^1(t) = f\left(\tilde{x}^1(t)\right) + \sum_{i=1}^{m} g_i\left(\tilde{x}^1(t)\right) u_i(t), \tag{5.211}$$

$$u_1(t) \in U_1, \tag{5.212}$$

$$u_j(t) = \tilde{u}_j^*(t|t_k), \quad j = 2, \ldots, m, \tag{5.213}$$

$$\tilde{x}(t_k) = x(t_k), \tag{5.214}$$

$$\frac{\partial V(x(t_k))}{\partial x} g_1\left(x(t_k)\right) u_1(t_k) \leq \frac{\partial V(x(t_k))}{\partial x} g_1\left(x(t_k)\right) h_1\left(x(t_k)\right). \tag{5.215}$$

In the LMPC 1 formulation of Eqs. 5.210–5.215, LMPC 1 takes advantage of the knowledge of $m - 1$ feasibility problems (i.e., $\tilde{u}_j^*$, $j = 2, \ldots, m$) to predict the future evolution of the system $\tilde{x}^1$. Let $u_1^*(t|t_k)$ denote the optimal solution of the optimization problem of Eqs. 5.210–5.215.

Based on the solutions of the m LMPC optimization problems, the manipulated inputs of the DMPC design are defined as follows:

$$u_i(t) = u_i^*(t|t_k), \quad \forall t \in [t_k, t_{k+1}), i = 1, \ldots, m. \tag{5.216}$$

Remark 5.17 Note that the white gaussian noise considered in this section is the accumulation of thermal effects and quantization errors. We do not consider the effects of multi-path transmission, terrain blocking, interference, etc. Furthermore, we assume that when package loss happens, all of the information that should be transmitted is lost; however, without loss of generality, the method presented in this section can be extended to the case in which data loss happens only in some packets of information following a similar methodology like Eqs. 5.206–5.209 to deal with this issue. The interested reader may refer to [15, 87] for more details on communication channel modeling.

5.7.3 Stability Properties

As it will be proved in Theorem 5.6 below, the DMPC framework takes advantage of the constraints of Eqs. 5.204 and 5.215 to compute the optimal trajectories $u_1, \ldots, u_m$ such that the Lyapunov function value $V(x(t_k))$ is a decreasing sequence with a lower bound and achieves the closed-loop stability of the system.

Theorem 5.6 *Consider the system of Eq. 5.1 in closed-loop under the DMPC design of Eqs. 5.199–5.216 based on a control law $h(x)$ that satisfies the conditions of Eqs. 5.5–5.8. Let $\varepsilon_w > 0$, $\Delta > 0$ and $\rho > \rho_s > 0$ satisfy the following constraint:*

$$-\alpha_3\left(\alpha_2^{-1}(\rho_s)\right) + \left(L_x' + \sum_{i=1}^{m} L_{u_i}' u_i^{\max}\right) M\Delta + L_w'\theta \leq -\varepsilon_w/\Delta. \tag{5.217}$$

If $x(t_0) \in \Omega_\rho$ and if $\rho_{\min} \leq \rho$ where $\rho_{\min} = \max\{V(x(t+\Delta)) : V(x(t)) \leq \rho_s\}$, then the state $x(t)$ of the closed-loop system is ultimately bounded in $\Omega_{\rho_{\min}}$.

Proof The proof consists of two parts. We first prove that the optimization problems of Eqs. 5.199–5.204 and 5.210–5.215 are feasible for all states $x \in \Omega_\rho$. Subsequently, we prove that, under the DMPC design of Eqs. 5.199–5.216, the state of the system of Eq. 5.1 is ultimately bounded in a region that contains the origin.

Part 1: First, we consider the feasibility of LMPC j of Eqs. 5.199–5.204 ($j = 2, \ldots, m$) and of LMPC 1 of Eqs. 5.210–5.215. All input trajectories of $u_j(t)$ ($j = 1, \ldots, m$) such that $u_j(t) = u_{n,j}(t|t_k)$, $\forall t \in [t_k, t_{k+N})$ satisfy all the constraints (including the input constraints of Eqs. 5.203 and 5.212 and the constraints of Eq. 5.204 and 5.215) of LMPC j, thus the feasibility of LMPC j as well as LMPC 1 is obtained.

Part 2: Considering the inequality of Eq. 5.6, addition of inequalities of Eqs. 5.204 for $j = 2, \ldots, m$ and 5.215 implies that if $x(t_k) \in \Omega_\rho$, the following inequality holds:

$$\frac{\partial V(x(t_k))}{\partial x}\left(f(x(t_k)) + \sum_{i=1}^{m} g_i(x(t_k))u_i^*(t_k|t_k)\right)$$
$$\leq \frac{\partial V(x(t_k))}{\partial x}\left(f(x(t_k)) + \sum_{i=1}^{m} g_i(x(t_k))u_{n,i}(t_k|t_k)\right)$$
$$\leq -\alpha_3(\|x(t_k)\|). \tag{5.218}$$

The time derivative of the Lyapunov function along the state trajectory $x(t)$ of the system of Eq. 5.1 in $t \in [t_k, t_{k+1})$ is given by:

$$\dot{V}(x(t)) = \frac{\partial V(x)}{\partial x}\left(f(x(t)) + \sum_{i=1}^{m} g_i(x(t))u_i^*(t_k|t_k) + k(x(t))w(t)\right). \tag{5.219}$$

Adding and subtracting $\frac{\partial V(x(t_k))}{\partial x}(f(x(t_k)) + \sum_{i=1}^{m} g_i(x(t_k))u_i^*(t_k|t_k))$ to the right-hand side of Eq. 5.219 and taking Eq. 5.218 into account, we obtain the following inequality:

$$\dot{V}(x(t)) \leq -\alpha_3(\|x(t_k)\|)$$
$$+ \frac{\partial V(x)}{\partial x}\left(f(x(t)) + \sum_{i=1}^{m} g_i(x(t))u_i^*(t_k|t_k) + k(x(t))w(t)\right)$$
$$- \frac{\partial V(x(t_k))}{\partial x}\left(f(x(t_k)) + \sum_{i=1}^{m} g_i(x(t_k))u_i^*(t_k|t_k)\right). \tag{5.220}$$

From Eq. 5.5, Eqs. 5.14–5.16, and the inequality of Eq. 5.220, the following inequality is obtained for all $x(t_k) \in \Omega_\rho / \Omega_{\rho_s}$:

$$\dot{V}(x(t)) \leq -\alpha_3\left(\alpha_2^{-1}(\rho_s)\right) + L'_w \|w(t)\| + \left(L'_x + \sum_{i=1}^{m} L'_{u_i} u_i^*(t_k|t_k)\right)\|x(t) - x(t_k)\|.$$

(5.221)

Taking into account Eq. 5.9 and the continuity of $x(t)$, the following bound can be written for all $t \in [t_k, t_{k+1})$, $\|x(t) - x(t_k)\| \leq M\Delta$. Using this expression, we obtain the following bound on the time derivative of the Lyapunov function for $t \in [t_k, t_{k+1})$, for all initial states $x(t_k) \in \Omega_\rho / \Omega_{\rho_s}$:

$$\dot{V}(x(t)) \leq -\alpha_3\left(\alpha_2^{-1}(\rho_s)\right) + \left(L'_x + \sum_{i=1}^{m} L'_{u_i} u_i^{\max}\right)M\Delta + L'_w\theta.$$

(5.222)

If the condition of Eq. 5.217 is satisfied, then there exists $\varepsilon_w > 0$ such that the following inequality holds for $x(t_k) \in \Omega_\rho / \Omega_{\rho_s}$:

$$\dot{V}(x(t)) \leq -\varepsilon_w / \Delta, \quad \forall t \in [t_k, t_{k+1}).$$

(5.223)

Integrating this bound on $t \in [t_k, t_{k+1})$, we obtain that:

$$V(x(t_{k+1})) \leq V(x(t_k)) - \varepsilon_w,$$

(5.224)

$$V(x(t)) \leq V(x(t_k)), \quad \forall t \in [t_k, t_{k+1})$$

(5.225)

for all $x(t_k) \in \Omega_\rho / \Omega_{\rho_s}$. Using Eqs. 5.224–5.225 recursively, it is proved that, if $x(t_0) \in \Omega_\rho / \Omega_{\rho_s}$, the state converges to Ω_{ρ_s} in a finite number of sampling times without leaving the stability region. Once the state converges to $\Omega_{\rho_s} \subseteq \Omega_{\rho_{\min}}$, it remains inside $\Omega_{\rho_{\min}}$ for all times. This statement holds because of the definition of $\rho_{\min}$. This proves that the closed-loop system under the DMPC of Eqs. 5.199–5.216 is ultimately bounded in $\Omega_{\rho_{\min}}$.   □

Remark 5.18 Note that the use of the corrupted input trajectory information of u_j (i.e., $\tilde{u}_j$) where $j = 2, \ldots, m$ does not affect the feasibility of the optimization problems of Eqs. 5.199–5.204 and 5.210–5.215 as well as the stability of the closed-loop system; however, it does affect the closed-loop system performance. This is the reason for the introduction of the feasibility problem of Eqs. 5.206–5.209 which is used to decide whether the corrupted information can be used or not to improve the closed-loop performance. An application of the DMPC architecture with the feasibility problem of Eqs. 5.199–5.216 to a chemical process can be found in [30].

5.8 Conclusions

In this chapter, we designed sequential and iterative DMPC schemes for large-scale nonlinear systems. In the sequential DMPC architecture, the distributed controllers adopt a one-directional communication strategy and are evaluated in sequence and once at each sampling time; in the iterative DMPC architecture, the distributed controllers utilize a bidirectional communication strategy, are evaluated in parallel and iterate to improve closed-loop performance. We considered three cases for the design of the sequential and iterative DMPC schemes: systems with continuous, synchronous state measurements, systems with asynchronous measurements and systems with delayed measurements. For all the three cases, appropriate implementation strategies, suitable Lyapunov-based stability constraints and sufficient conditions under which practical closed-loop stability is ensured, were provided. Extensive simulations using a catalytic alkylation of benzene process example were carried out to compare the DMPC architectures with existing centralized LMPC algorithms from computational time and closed-loop performance points of view. Moreover, we focused on a hierarchical type DMPC and discussed how to handle communication disruptions in the communication between the distributed controllers by incorporating feasibility problems to decide the reliability of the transmitted information.

Chapter 6
Multirate Distributed Model Predictive Control

6.1 Introduction

In Chap. 5, we considered the design of DMPC architectures for large-scale nonlinear systems assuming that all the measurements of the system states are available at the same sampling instants. In this chapter, we consider the design of a network-based DMPC system using multirate sampling for large-scale nonlinear uncertain systems composed of several coupled subsystems. Specifically, we assume that the states of each local subsystem can be divided into fast sampled states (which are available every sampling time) and slowly sampled states (which are available every several sampling times). The distributed model predictive controllers are connected through a shared communication network and cooperate in an iterative fashion at time instants in which full system state measurements (both fast and slow) are available, to guarantee closed-loop stability. When only local subsystem fast sampled state information is available, the distributed controllers operate in a decentralized fashion to improve closed-loop performance. In this control architecture, the controllers are designed via LMPC techniques taking into account bounded measurement noise, process disturbances and communication noise. Sufficient conditions under which the state of the closed-loop system is ultimately bounded in an invariant region containing the origin are derived. The theoretical results are demonstrated through a nonlinear chemical process example. The results of this chapter were first presented in [31].

6.2 System Description

We consider a class of nonlinear systems which is composed of m interconnected subsystems where each of the subsystems can be described by the following state-space model:

$$\dot{x}_i(t) = f_i(x) + g_{si}(x)u_i(t) + k_i(x)w_i(t), \tag{6.1}$$

P.D. Christofides et al., *Networked and Distributed Predictive Control*,
Advances in Industrial Control,
DOI 10.1007/978-0-85729-582-8_6, © Springer-Verlag London Limited 2011

where $i = 1, \ldots, m$, $x_i(t) \in R^{n_i}$ denotes the vector of state variables of subsystem i, $u_i(t) \in R^{m_i}$ and $w_i(t) \in R^{w_i}$ denote the set of control (manipulated) inputs and disturbances associated with subsystem i, respectively. The variable $x \in R^n$ denotes the state of the entire nonlinear system which is composed of the states of the m subsystems, that is:

$$x = \left[x_1^T \cdots x_i^T \cdots x_m^T \right]^T \in R^n. \tag{6.2}$$

The dynamics of x can be described in a compact form as follows:

$$\dot{x}(t) = f(x) + \sum_{i=1}^{m} g_i(x) u_i(t) + k(x) w(t), \tag{6.3}$$

where $f = [f_1^T \cdots f_i^T \cdots f_m^T]^T$, $g_i = [0^T \cdots g_{si}^T \cdots 0^T]^T$ with 0 being the zero matrix of appropriate dimensions, k is a matrix composed of k_i $(i = 1, \ldots, m)$ and zeros whose explicit expression is omitted for brevity, and $w = [w_1^T \cdots w_i^T \cdots w_m^T]^T$ is assumed to be bounded, that is:

$$W := \left\{ w \in R^w : \|w\| \le \theta, \theta > 0 \right\} \tag{6.4}$$

with θ being a known positive real number.

The m sets of inputs are restricted to be in m nonempty convex sets $U_i \subseteq R^{m u_i}$, $i = 1, \ldots, m$, which are defined as:

$$U_i := \left\{ u_i \in R^{m_i} : \|u_i\| \le u_i^{\max} \right\}, \tag{6.5}$$

where $u_i^{\max}$, $i = 1, \ldots, m$, are the magnitudes of the input constraints in an element-wise manner. We will design m controllers to compute the m sets of control inputs u_i, $i = 1, \ldots, m$, respectively. We will refer to the controller computing u_i associated with subsystem i as controller i.

We assume that f, g_i, $i = 1, \ldots, m$, and k are locally Lipschitz vector, matrix and matrix functions, respectively, and that the origin is an equilibrium point of the unforced nominal system (i.e., system of Eq. 6.3 with $u_i(t) = 0$, $i = 1, \ldots, m$, $w(t) = 0$ for all t) which implies that $f(0) = 0$.

6.3 Modeling of Measurements and Communication Networks

We assume that the states of each of the m subsystems, x_i $(i = 1, \ldots, m)$, are divided into two parts: $x_{f,i}$, states that can be measured at each sampling time (e.g., temperatures and pressures) and $x_{s,i}$, states which are sampled at a relatively slow rate (e.g., species concentrations). Specifically, we assume that $x_{f,i}$, are available at synchronous time instants $t_p = t_0 + p\Delta$, $p = 0, 1, \ldots$, where t_0 is the initial time and Δ is the sampling time; and assume that $x_{s,i}$, are available every T sampling times (i.e., $x_{s,i}$, are available at t_k with $k = 0, T, 2T, \ldots$). Note that, in order to simplify the development, we assume that the slowly sampled states of different subsystems are all available at the same time instants. This modeling of measurements is relevant to systems involving heterogeneous measurements which have different sampling rates; please see the example in Sect. 6.6.

We also assume that for each subsystem its local sensors, actuators and controller are connected using point-to-point links, which implies that $x_{f,i}$ and $x_{s,i}$ are available without delay to controller i once they are measured. We further assume that the controllers for different subsystems are connected through a shared communication network and communicate when the full system state is available (i.e., at time instants t_k with $k = 0, T, 2T, \ldots$). When each predictive controller communicates with the rest of the controllers, they share state and future input trajectories information.

Moreover, in addition to process disturbances, we consider measurement noise and communication network noise. Specifically, we consider measurement noise caused by the lack of complete accuracy of measurement sensors. This type of noise is defined as the difference between the reading value of a state from a sensor and the true value of the state. We assume that the sensor reading values of states $x_{f,i}$ and $x_{s,i}$ are $\check{x}_{f,i}^s$ and $\check{x}_{s,i}^s$, respectively; and $\check{x}_{f,i}^s$ and $\check{x}_{s,i}^s$ are modeled as follows:

$$\check{x}_{f,i}^s = x_{f,i} + n_{x_{f,i}}^s, \tag{6.6}$$

$$\check{x}_{s,i}^s = x_{s,i} + n_{x_{s,i}}^s, \tag{6.7}$$

where $n_{x_{f,i}}^s$ and $n_{x_{s,i}}^s$ are the measurement noise terms associated with $x_{f,i}$ and $x_{s,i}$, respectively. The measurement noise is assumed to be bounded, that is, $\|n_{x_{f,i}}^s\| \leq \theta_{x_{f,i}}^s$ and $\|n_{x_{s,i}}^s\| \leq \theta_{x_{s,i}}^s$ with $\theta_{x_{f,i}}^s$ and $\theta_{x_{s,i}}^s$ being positive real numbers. It should be mentioned that this assumption on the type of measurement noise is meaningful from a practical standpoint due to the limit on the accuracy of the measurement sensors and the fact that measurement noise is usually modeled as a percentage of the actual value.

In addition to measurement noise, we consider communication channel noise of the shared communication network. At t_k with $k = 0, T, 2T, \ldots$, when fast and slowly sampled states are available to each controller, the distributed controllers exchange information which is subject to communication channel noise. Specifically, we assume that controller i sends $\check{x}_i^s = [\check{x}_{f,i}^{s,T} \ \check{x}_{s,i}^{s,T}]^T$ as well as its control input trajectory u_i to the other controllers; and the values received by controller j ($j \neq i$), $\check{x}_i^j$ and $\check{u}_i^j$, are modeled as follows:

$$\check{x}_i^j = \check{x}_i^s + n_{x_i}^{c,j}, \tag{6.8}$$

$$\check{u}_i^j = u_i + n_{u_i}^j, \tag{6.9}$$

where $n_{x_i}^{c,j}$ and $n_{u_i}^j$ are the communication noise terms. The communication noise terms are also assumed to be bounded; that is, $\|n_{x_i}^{c,j}\| \leq \theta_{x_i}^{c,j}$ and $\|n_{u_i}^j\| \leq \theta_{u_i}$ with $\theta_{x_i}^{c,j}$ and θ_{u_i} being positive real numbers. The noise terms in Eqs. 6.8–6.9 are gaussian white noise variables with zero mean and covariance matrix $\sigma_{x_i^j}^2$ and $\sigma_{u_i^j}^2$ with appropriate dimensions, respectively. The power of the input signal (which can be system state variables or control input trajectories) should be much greater than the channel noise variance in a way such that the input signal does not disappear in the noise, which implies that $\sigma^2_{x^j_i}$ and $\sigma^2_{u_i}$ should be sufficiently small. Furthermore, in order to get deterministic stability results, we assume that the channel noise is also bounded by $\theta^{c,j}_{x_i}$ and θ_{u_i}.

According to the above modeling, at time t_k with $k = 0, T, 2T, \ldots$ when fast and slowly sampled states are available, the state information received by controller i ($i = 1, \ldots, m$) is described as follows:

$$\check{x}^i(t_k) = \left[\check{x}^i_1, \ldots, \check{x}^i_{i-1}, \check{x}^s_i, \check{x}^i_{i+1}, \ldots, \check{x}^i_m\right] = x(t_k) + n^i_x, \qquad (6.10)$$

where $n^i_x \in R^{n_x}$ denotes combined communication and measurement noise and $\|n^i_x\| \le \theta^i_x$ with θ^i_x being a suitable composition of $\theta^s_{x_{f,i}}$, $\theta^s_{x_{s,i}}$ and $\theta^{c,i}_{x_j}$ ($j \ne i$).

This class of systems is relevant to the case of large-scale chemical processes that are controlled by distributed control systems that exchange information over a shared communication network through which it is not cost-effective to communicate at every sampling time. Instead, in order to achieve closed-loop stability and good closed-loop performance, the controllers communicate every several sampling times. Please see Fig. 6.1 in Sect. 6.5 for a schematic of such type of DMPC system with the distributed controllers designed via LMPC techniques.

6.4 Lyapunov-Based Control

We assume that there exists a locally Lipschitz nonlinear control law $h(x) = [h_1(x)^T \cdots h_m(x)^T]^T$ with $u_i = h_i(x)$, $i = 1, \ldots, m$, which renders the origin of the nominal interconnected closed-loop system asymptotically stable while satisfying the input constraints for all x inside the closed-loop stability region. This assumption implies that there exist class $\mathcal{K}$ functions $\alpha_i(\cdot)$, $i = 1, 2, 3, 4$ and a continuously differentiable Lyapunov function $V(x)$ for the nominal closed-loop system, that satisfy the following inequalities:

$$\alpha_1(\|x\|) \le V(x) \le \alpha_2(\|x\|), \qquad (6.11)$$

$$\frac{\partial V(x)}{\partial x}\left(f(x) + \sum_{i=1}^{m} g_i(x)h_i(x)\right) \le -\alpha_3(\|x\|), \qquad (6.12)$$

$$\left\|\frac{\partial V(x)}{\partial x}\right\| \le \alpha_4(\|x\|), \qquad (6.13)$$

$$h_i(x) \in U_i, \quad i = 1, \ldots, m \qquad (6.14)$$

for all $x \in O \subseteq R^{n_x}$ where O is an open neighborhood of the origin. We denote the region $\Omega_\rho \subseteq O$ as the stability region of the closed-loop system under $h(x)$.

By continuity, the local Lipschitz property assumed for the vector fields $f(x)$, $g_i(x)$, $i = 1, \ldots, m$, $k(x)$ and $h(x)$ and taking into account that the manipulated

Fig. 6.1 DMPC architecture
with multirate sampling
(*solid line* denotes fast state
sampling and/or
point-to-point links; *dashed
line* denotes slow state
sampling and/or shared
communication networks)

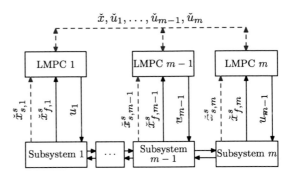

inputs u_i, $i = 1, \ldots, m$, and the disturbance w are bounded in convex sets, there exist positive constants M, M_{g_i}, L_x, L_{u_i}, L_{h_i} and L_w $(i = 1, \ldots, m)$ such that:

$$\left\| f(x) + \sum_{i=1}^{m} g_i(x)u_i + k(x)w \right\| \leq M, \tag{6.15}$$

$$\|g_i(x)\| \leq M_{g_i}, \quad i = 1, \ldots, m, \tag{6.16}$$

$$\|f(x) - f(x')\| \leq L_x \|x - x'\|, \tag{6.17}$$

$$\|g_i(x) - g_i(x')\| \leq L_{u_i} \|x - x'\|, \quad i = 1, \ldots, m, \tag{6.18}$$

$$\|h_i(x) - h_i(x')\| \leq L_{h_i} \|x - x'\|, \quad i = 1, \ldots, m, \tag{6.19}$$

$$\|k(x)\| \leq L_w \tag{6.20}$$

for all $x, x' \in \Omega_\rho$, $u_i \in U_i$, $i = 1, \ldots, m$, and $w \in W$. In addition, by the continuous differentiable property of the Lyapunov function $V(x)$, there exist positive constants L'_x, L'_{u_i}, C_{g_i}, $i = 1, \ldots, m$, and L'_w such that:

$$\left\| \frac{\partial V(x)}{\partial x} f(x) - \frac{\partial V(x')}{\partial x} f(x') \right\| \leq L'_x \|x - x'\|, \tag{6.21}$$

$$\left\| \frac{\partial V(x)}{\partial x} g_i(x) - \frac{\partial V(x')}{\partial x} g_i(x') \right\| \leq L'_{u_i} \|x - x'\|, \quad i = 1, \ldots, m, \tag{6.22}$$

$$\left\| \frac{\partial V(x)}{\partial x} g_i(x) \right\| \leq C_{g_i}, \quad i = 1, \ldots, m, \tag{6.23}$$

$$\left\| \frac{\partial V(x)}{\partial x} k(x) \right\| \leq L'_w \tag{6.24}$$

for all $x, x' \in \Omega_\rho$, $u_i \in U_i$, $i = 1, \ldots, m$, and $w \in W$.

6.5 Multirate DMPC

In this section, the m controllers manipulating the m sets of inputs will be designed through LMPC techniques. For the LMPC associated with controller i, $i = 1, \ldots, m$, we will refer to it as LMPC i. A schematic of the control system is shown in Fig. 6.1.

6.5.1 Multirate DMPC Formulation

At a sampling time in which slowly and fast sampled states are available, the distributed controllers coordinate their actions and predict future input trajectories which, if applied until the next instant that both slowly and fast sampled states are available, guarantee closed-loop stability. At a sampling time in which only fast sampled states are available, each distributed controller tries to further optimize the input trajectories calculated at the last instant in which the controllers communicated, within a constrained set of values to improve the closed-loop performance with the help of the available fast sampled states of its subsystem.

We propose to adopt an iterative DMPC approach when we have access to both fast and slowly sampled state measurements. Specifically, the implementation strategy of the DMPC architecture at time instants in which fast and slowly sampled states are available is as follows:

1. At t_k with $k = 0, T, 2T, \ldots$, all the controllers first broadcast their local subsystem states to the other controllers and then evaluate their future input trajectories in an iterative fashion with initial input guesses generated by $h(\cdot)$.
2. At iteration c $(c \geq 1)$
    2.1. Each controller evaluates its own future input trajectory based on $\check{x}^i(t_k)$ (noisy version of $x(t_k)$) and the last received control input trajectories (when $c = 1$, initial input guesses generated by $h(\cdot)$ are used).
    2.2. All the distributed controllers exchange their latest future input trajectories. Based on the input information, each controller calculates and stores the corresponding value of the cost function.
3. If a termination condition is satisfied, each controller sends its entire future input trajectory corresponding to the smallest value of the cost to its actuators; if the termination condition is not satisfied, go to Step 2 $(c \leftarrow c + 1)$.

The implementation strategy of the DMPC architecture at time instants when only local fast sampled states are available is as follows:

1. Controller i, $i = 1, \ldots, m$, receives its local fast sampled states, $\check{x}^s_{f,i}$ which are affected by measurement noise.
2. Each controller i estimates the current full system state and evaluates its future input trajectory and sends the first step input value to its actuators.

In the sequel, we describe these steps in detail.

We first describe the design of the LMPCs at time instants in which fast and slowly sampled states are available. To proceed, we define a nominal sampled trajectory for each subsystem $x_h^i(t|t_k)$, $k = 0, T, 2T, \ldots$, which will be employed in the construction of the stability constraint of LMPC i ($i = 1, \ldots, m$). This nominal sampled trajectory is obtained by integrating recursively, for $t \in [t_k, t_{k+T})$ and $k = 0, T, 2T, \ldots$, the following equation:

$$\dot{x}_h^i(t|t_k) = f\left(x_h^i(t|t_k)\right) + \sum_{i=1}^m g_i\left(x_h^i(t|t_k)\right) h_i\left(x_h^i(t_{k+l}|t_k)\right), \quad \forall \in [t_{k+l}, t_{k+l+1}),$$

(6.25)

$$x_h^i(t_k|t_k) = \check{x}^i(t_k),$$

(6.26)

where $l = 0, \ldots, T - 1$, $\check{x}^i(t_k)$ is the system state received by controller i at t_k. Based on this sampled trajectory, we can define the following input trajectories:

$$u_{h,j}^i(t|t_k) = h_j\left(x_h^i(t_{k+l}|t_k)\right), \quad j = 1, \ldots, m, \forall t \in [t_{k+l}, t_{k+l+1}), l = 0, \ldots, T - 1.$$

(6.27)

From the definition of x_h^i, we see that this trajectory is a prediction of the evolution of the system of Eq. 6.3 under the control law $h(x)$ applied in a sample-and-hold fashion. This sampled trajectory, $x_h^i(t|t_k)$, will be used in the formulation of LMPC i.

At time t_k, $k = 0, T, 2T, \ldots$, the LMPCs are evaluated in an iterative fashion to obtain the future input trajectories. Specifically, the optimization problem of LMPC j at iteration c is as follows:

$$\min_{u_j \in S(\Delta)} \int_{t_k}^{t_k+N} \left[\left\|\tilde{x}^j(\tau)\right\|_{Q_c} + \sum_{i=1}^m \left\|u_i(\tau)\right\|_{R_{ci}}\right] d\tau,$$

(6.28)

$$\text{s.t.} \quad \dot{\tilde{x}}^j(t) = f\left(\tilde{x}^j(t)\right) + \sum_{i=1}^m g_i\left(\tilde{x}^j(t)\right) u_i(t),$$

(6.29)

$$u_i(t) = \check{u}_i^{*,c-1}(t|t_k), \quad \forall i, \neq j,$$

(6.30)

$$\left\|u_j(t) - u_j^{*,c-1}(t|t_k)\right\| \leq \Delta u_j, \quad \forall t \in [t_k, t_{k+T}),$$

(6.31)

$$u_j(t) \in U_j,$$

(6.32)

$$\tilde{x}^j(t_k) = \check{x}^j(t_k),$$

(6.33)

$$\frac{\partial V(\tilde{x}^j(t))}{\partial x}\left(\frac{1}{m} f\left(\tilde{x}^j(t)\right) + g_j\left(\tilde{x}^j(t)\right) u_j(t)\right)$$

$$\leq \frac{\partial V(x_h^j(t|t_k))}{\partial x}\left(\frac{1}{m} f\left(x_h^j(t|t_k)\right) + g_j\left(x_h^j(t|t_k)\right) u_{h,j}^j(t|t_k)\right),$$

$$\forall t \in [t_k, t_{k+T}),$$

(6.34)

where state $\tilde{x}^j$ is the predicted trajectory of the nominal system with u_j computed by this LMPC and all the other inputs are received from the other controllers (i.e., $\breve{u}_i^{*,c-1}(t|t_k)$ which is a noisy version of $u_i^{*,c-1}(t|t_k)$).

The optimal solution to this optimization problem is denoted by $u_j^{*,c}(t|t_k)$ which is defined for $t \in [t_k, t_{k+N})$. Accordingly, we define the final optimal input trajectory of LMPC j (that is, the optimal trajectories computed at the last iteration) as $u_j^{*,f}(t|t_k)$ which is also defined for $t \in [t_k, t_{k+N})$. Note that for the first iteration of each distributed LMPC, the input trajectories defined in Eq. 6.27 are used as the initial input trajectory guesses; that is, $u_i^{*,0} = u_{h,i}^j$ with $i = 1, \ldots, m$. The constraint of Eq. 6.34 is used to guarantee the closed-loop stability and the prediction horizon N should be chosen to satisfy $N \geq T$.

The manipulated inputs of the above control design from time t_k to t_{k+1} ($k = 0, T, 2T, \ldots$) are defined as follows:

$$u_i(t) = u_i^{*,f}(t|t_k), \quad i = 1, \ldots, m, \ \forall t \in [t_k, t_{k+1}). \tag{6.35}$$

Next, we describe the design of the distributed controllers at the time instants in which only local fast sampled states are available. In order to improve the performance, between two slow sampling times, each controller uses the available local fast sampled measurements to adjust its control input based on the calculated optimal input trajectory for the current time obtained at the last time instant in which fast and slowly sampled states were available. In order to guarantee closed-loop stability, the maximum deviation of the adjusted inputs from the optimal input trajectory at each time step is bounded.

Between two slow sampling times, each controller estimates the current full system state using an observer based on the system model and the available information. Specifically, the observer for controller i takes the following form for $t \in [t_{l-1}, t_l)$:

$$\dot{\hat{x}}^i(t) = f(\hat{x}^i(t)) + g_i(\hat{x}^i(t))u_i^*(t) + \sum_{j=1}^{m, j \neq i} g_j(\hat{x}^i(t))\breve{u}_j^{*,i}(t|t_k), \tag{6.36}$$

$$\hat{x}^i(t_{l-1}) = x_e^i(t_{l-1}), \tag{6.37}$$

where $\hat{x}^i$ is the state of this observer, $\breve{u}_j^{*,i}(t|t_k)$ is the optimal input trajectory of LMPC j ($j = 1, \ldots, m, \ j \neq i$) received by LMPC i (i.e., it is a noisy version of $u_j^{*,f}(t|t_k) + n_{u_j}^i$), $u_i^*(t)$ is the actual input that has been applied to subsystem i, and $x_e^i(t_{l-1})$ is the full state estimate obtained at t_{l-1}. The state estimate $x_e^i(t_l)$, $l \neq 0, T, 2T, \ldots$, is a combination of the state of the observer of Eqs. 6.36–6.37 and of the available local state information $\breve{x}_{f,i}^s(t_l)$ as follows:

$$x_e^i(t_l) = \left[\hat{x}_1^i(t_l)^T \cdots \breve{x}_i(t_l)^T \cdots \hat{x}_m^i(t_l)^T\right]^T, \tag{6.38}$$

where $\breve{x}_i(t_l)^T = [\breve{x}_{f,i}^{s,T} \ \hat{x}_{s,i}^T]$.

The optimization problem of LMPC j for a time instant t_l, $l \neq 0, T, 2T, \ldots$ is as follows:

$$\min_{u_j \in S(\Delta)} \int_{t_k}^{t_k+N} \left[\left\| \tilde{x}^j(\tau) \right\|_{Q_c} + \sum_{i=1}^{m} \left\| u_i(\tau) \right\|_{R_{ci}} \right] d\tau, \tag{6.39}$$

$$\text{s.t.} \quad \dot{\tilde{x}}^j(t) = f\left(\tilde{x}^j(t)\right) + \sum_{i=1}^{m} g_i\left(\tilde{x}^j(t)\right) u_i(t), \tag{6.40}$$

$$u_i(t) = \breve{u}_i^{*,j}(t|t_k), \quad \forall i \neq j, t \in [t_l, t_{k+N}), \tag{6.41}$$

$$u_i(t) = h_i(\tilde{x}^j(t)), \quad \forall i \neq j, t \in [t_{k+N}, t_{l+N}), \tag{6.42}$$

$$\left\| u_j(t) - u_j^{*,f}(t|t_k) \right\| \leq \Delta u_j, \quad t \in [t_l, t_{k+N}), \tag{6.43}$$

$$u_j(t) \in U_j, \tag{6.44}$$

$$\tilde{x}^j(t_l) = x_e^j(t_l), \tag{6.45}$$

where t_k is the last time instant in which both fast and slowly sampled states are available, the state $\tilde{x}^j$ is the predicted trajectory of the nominal system with u_j computed by this LMPC and all the other inputs are determined by the constraints of Eqs. 6.41 and 6.42. In this optimization problem, the input u_j is restricted to be within a bounded region around the reference input trajectories given by $u_j^{*,f}(t|t_k)$ and $h(x)$. This ensures that the stability of the closed-loop system is maintained; please also see the proof of Theorem 6.1 below. The optimal solution to this optimization problem is denoted by $u_j^{*,l}(t|t_l)$ which is defined for $t \in [t_l, t_{l+N})$.

The manipulated inputs of the control design of Eqs. 6.39–6.45 from t_l to t_{l+1} ($l \neq 0, T, 2T, \ldots$) are defined as follows:

$$u_i(t) = u_i^{*,l}(t|t_l), \quad i = 1, \ldots, m, \forall t \in [t_l, t_{l+1}). \tag{6.46}$$

In the design of Eqs. 6.28–6.35 and 6.39–6.46, the closed-loop stability of the system of Eq. 6.3 is guaranteed by the design of Eqs. 6.28–6.35 at each sampling time t_k, $k = 0, T, 2T, \ldots$, when the full state measurements are available. The design of Eqs. 6.39–6.46 takes advantage of the predicted input trajectories $u_i^{*,f}$, $i = 1, \ldots, m$, at sampling times t_k, $k = 0, T, 2T, \ldots$, and the additional available fast-sampling state measurements to adjust the predicted inputs, $u_i^{*,f}$, to improve the closed-loop performance.

Remark 6.1 In the proposed implementation strategies, we adopt an iterative DMPC approach when both the fast and slowly sampled states are available. Note that a sequential DMPC approach or a centralized LMPC approach can also be applied in the proposed multirate DMPC framework.

6.5.2 Stability Properties

The multirate DMPC of Eqs. 6.28–6.35 and 6.39–6.46 computes the inputs u_i, $i = 1, \ldots, m$, applied to the system of Eq. 6.3 in a way such that in the closed-loop system, the value of the Lyapunov function at time instant t_k (i.e., $V(x(t_k))$) is a decreasing sequence of values with a lower bound. This is achieved due to the constraints of Eq. 6.34 incorporated in each LMPC. This property is presented in Theorem 6.1 below. To prove this theorem, we need the following definitions, propositions and corollaries.

Definition 6.1 We define Ω_{ρ_n} as follows:

$$\rho_n = \max\{V(x(t)) : (x(t) + n) \in \Omega_\rho, \|n\| \le \theta_x\}, \tag{6.47}$$

where $\theta_x = \max_{1 \le i \le m}\{\theta_x^i\}$ defines the upper bound on the noise n. The region Ω_{ρ_n} will be used as the stability region of the system under the control law $h(x)$ in the presence of measurement noise, process disturbances and communication noise.

Definition 6.2 The closed-loop state trajectory of the nominal system for time $t \in [t_k, t_{k+1})$ under the nonlinear control law $h(x)$ based on actual system state $(x(t_k))$ and applied in sample and hold fashion is denoted by $x_{h,2}(t)$ which is obtained by integrating, for $t \in [t_k, t_{k+1})$, the following equation:

$$\dot{x}_{h,2}(t) = f\left(x_{h,2}(t)\right) + \sum_{i=1}^{m} g_i\left(x_{h,2}(t)\right)h_i\left(x_{h,2}(t_k)\right), \tag{6.48}$$

where $x_{h,2}(t_k) \in \Omega_{\rho_n}$.

Definition 6.3 The closed-loop state trajectory of the nominal system for time $t \in [t_k, t_{k+1})$ under the nonlinear control law $h(x)$ based on noisy system states and applied in sample and hold fashion is denoted by $x_h(t)$ which is obtained by integrating, for $t \in [t_k, t_{k+1})$, the following equation:

$$\dot{x}_h(t) = f\left(x_h(t)\right) + \sum_{i=1}^{m} g_i\left(x_h(t)\right)h_i\left(\check{x}_h(t_k)\right), \tag{6.49}$$

where $x_h(t_k) \in \Omega_{\rho_n}$ and $\check{x}_h(t_k) = x_{h,2}(t_k) + n(t_k)$, $\|n\| \le \theta_x$.

Proposition 6.1 *Consider the systems*:

$$\dot{x}_a(t) = f\left(x_a(t)\right) + \sum_{i=1}^{m} g_i\left(x_a(t)\right)h_i\left(\check{x}_a(0)\right), \tag{6.50}$$

$$\dot{x}_b(t) = f\left(x_b(t)\right) + \sum_{i=1}^{m} g_i\left(x_b(t)\right)h_i\left(\check{x}_b(0)\right), \tag{6.51}$$

where the initial states $x_a(0), x_b(0) \in \Omega_{\rho_n}$, $\|x_a(0) - x_b(0)\| \le \theta_{ab}$, $\|x_a(0) - \check{x}_a(0)\| \le \theta_a$ and $\|x_b(0) - \check{x}_b(0)\| \le \theta_b$. If $0 < \rho_n < \rho$, then there exists a function $f_E(\cdot, \cdot, \cdot, \cdot)$ such that:

$$\|x_a(t) - x_b(t)\| \le f_E(\theta_{ab}, \theta_a, \theta_b, t), \tag{6.52}$$

for all $x_a(t), x_b(t) \in \Omega_{\rho_n}$ with $f_E(\theta_{ab}, \theta_a, \theta_b, t) = (\theta_{ab} + \frac{L_2}{L_1})e^{L_1 t} - \frac{L_2}{L_1}$ where $L_1 = L_x + \sum_{i=1}^{m} u_i^{\max} L_{u_i}$, $L_2 = (\theta_a + \theta_b + \theta_{ab}) \sum_{i=1}^{m} M_{g_i} L_{h_i}$ with θ_{ab}, θ_a and θ_b being positive real numbers.

Proof Define the error vector as $e(t) = x_a(t) - x_b(t)$. The time derivative of the error, $e(t)$, is as follows:

$$\dot{e}(t) = f(x_a(t)) - f(x_b(t)) + \sum_{i=1}^{m} \left(g_i(x_a(t)) h_i(\check{x}_a(0)) - g_i(x_b(t)) h_i(\check{x}_b(0)) \right).$$
$$\tag{6.53}$$

Adding to and subtracting $\sum_{i=1}^{m} g_i(x_a(t)) h_i(\check{x}_b(0))$ from the above equation, we obtain:

$$\dot{e}(t) = f(x_a(t)) - f(x_b(t)) + \sum_{i=1}^{m} g_i(x_a(t)) \left(h_i(\check{x}_a(0)) - h_i(\check{x}_b(0)) \right)$$

$$+ \sum_{i=1}^{m} h_i(\check{x}_b(0)) \left(g_i(x_a(t)) - g_i(x_b(t)) \right). \tag{6.54}$$

Using the conditions defined in Eqs. 6.17–6.19 obtained by the local Lipschitz property assumed for the vector fields $f(\cdot)$, $g_i(\cdot)$, $h_i(\cdot)$, $i = 1, \ldots, m$, and the fact that $h_i(\cdot)$ satisfies input constraints ($h_i(\cdot) \in U_i$), we obtain the following inequality:

$$\|\dot{e}(t)\| \le L_x \|x_a(t) - x_b(t)\| + \sum_{i=1}^{m} M_{g_i} L_{h_i} \|\check{x}_a(0) - \check{x}_b(0)\|$$

$$+ \sum_{i=1}^{m} u_i^{\max} L_{u_i} \|x_a(t) - x_b(t)\|. \tag{6.55}$$

Using that $\|\check{x}_a(0) - \check{x}_b(0)\| \le \theta_a + \theta_b + \theta_{ab}$ and defining $L_1 = L_x + \sum_{i=1}^{m} u_i^{\max} L_{u_i}$ and $L_2 = (\theta_a + \theta_b + \theta_{ab}) \sum_{i=1}^{m} M_{g_i} L_{h_i}$, we obtain:

$$\|\dot{e}(t)\| \le L_1 \|e(t)\| + L_2. \tag{6.56}$$

Integrating $\|\dot{e}(t)\|$ with initial condition $\|e(0)\| \le \theta_{ab}$, the following bound on the norm of the error vector is obtained:

$$\|e(t)\| \le \left(\theta_{ab} + \frac{L_2}{L_1} \right) e^{L_1 t} - \frac{L_2}{L_1}. \tag{6.57}$$

This proves Proposition 6.1.   □

The following corollary which is an extension of Proposition 2.1 provides sufficient conditions that ensure that $h(\cdot)$ can achieve closed-loop stability of the nominal
system in the presence of bounded measurement and communication noise.

Corollary 6.1 *Consider the closed-loop nominal sampled trajectory $x_h(t)$ of the
system of Eq. 6.3 as defined in Definition 6.3. Let $\Delta, \varepsilon_s, \theta_x > 0$ and $0 < \rho_s < \rho_n < \rho$
satisfy:*

$$-\alpha_3\left(\alpha_2^{-1}(\rho_s)\right) + \left(L'_x + \sum_{i=1}^{m} u_i^{\max} L'_{u_i}\right)\left(f_E(0,0,\theta_x,\Delta) + M\Delta\right) + \theta_x \sum_{i=1}^{m} C_{g_i} L_{h_i}$$

$$\leq -\varepsilon_s/\Delta, \tag{6.58}$$

*where f_E is defined in Proposition 6.1. Then, for any k, if $x_h(t_k) \in \Omega_{\rho_n}/\Omega_{\rho_s}$ the
following inequalities hold:*

$$V\left(x_h(t_{k+1})\right) \leq V\left(x_h(t_k)\right) - \varepsilon_s \tag{6.59}$$
$$V\left(x_h(t)\right) \leq V\left(x_h(t_k)\right), \quad \forall t \in [t_k, t_{k+1}). \tag{6.60}$$

Also, if $\rho_{\min} \leq \rho_n$ where:

$$\rho_{\min} = \max\left\{V\left(x_h(t+\Delta)\right) : V\left(x_h(t)\right) \leq \rho_s\right\} \tag{6.61}$$

and $x_h(t_0) \in \Omega_{\rho_n}$, the following inequalities hold:

$$V\left(x_h(t_k)\right) \leq \max\left\{V\left(x_h(t_0)\right) - k\varepsilon_s, \rho_{\min}\right\}, \tag{6.62}$$
$$V\left(x_h(t)\right) \leq \max\left\{V\left(x_h(t_k)\right), \rho_{\min}\right\}, \quad \forall t \in [t_k, t_{k+1}). \tag{6.63}$$

Proof Following Definition 6.3, the time derivative of the Lyapunov function along
the nominal sampled trajectory $x_h(t)$ of the system of Eq. 6.3 in $t \in [t_k, t_{k+1})$ is
given by:

$$\dot{V}\left(x_h(t)\right) = \frac{\partial V(x_h(t))}{\partial x}\left(f\left(x_h(t)\right) + \sum_{i=1}^{m} g_i\left(x_h(t)\right)h_i\left(\check{x}_h(t_k)\right)\right). \tag{6.64}$$

Adding and subtracting $\frac{\partial V(x_{h,2}(t_k))}{\partial x}(f(x_{h,2}(t_k)) + \sum_{i=1}^{m} g_i(x_{h,2}(t_k))h_i(x_{h,2}(t_k)))$ to
and from the above equation and taking into account the Eq. 6.12, we obtain:

$$\dot{V}\left(x_h(t)\right) \leq -\alpha_3\left(\|x_{h,2}(t_k)\|\right) + \frac{\partial V(x_h(t))}{\partial x}f\left(x_h(t)\right) - \frac{\partial V(x_{h,2}(t_k))}{\partial x}f\left(x_{h,2}(t_k)\right)$$

$$+ \frac{\partial V(x_h(t))}{\partial x}\sum_{i=1}^{m} g_i\left(x_h(t)\right)h_i\left(\check{x}_h(t_k)\right)$$

$$- \frac{\partial V(x_{h,2}(t_k))}{\partial x}\sum_{i=1}^{m} g_i\left(x_{h,2}(t_k)\right)h_i\left(x_{h,2}(t_k)\right). \tag{6.65}$$

Adding and subtracting $\frac{\partial V(x_h(t))}{\partial x}\sum_{i=1}^{m}g_i(x_h(t))h_i(x_{h,2}(t_k))$ to and from the above inequality, we have:

$$\dot{V}(x_h(t)) \leq -\alpha_3\left(\|x_{h,2}(t_k)\|\right) + \frac{\partial V(x_h(t))}{\partial x}f(x_h(t)) - \frac{\partial V(x_{h,2}(t_k))}{\partial x}f(x_{h,2}(t_k))$$

$$+ \frac{\partial V(x_h(t))}{\partial x}\sum_{i=1}^{m}g_i(x_h(t))h_i(\check{x}_h(t_k))$$

$$- \frac{\partial V(x_h(t))}{\partial x}\sum_{i=1}^{m}g_i(x_h(t))h_i(x_{h,2}(t_k))$$

$$+ \frac{\partial V(x_h(t))}{\partial x}\sum_{i=1}^{m}g_i(x_h(t))h_i(x_{h,2}(t_k))$$

$$- \frac{\partial V(x_{h,2}(t_k))}{\partial x}\sum_{i=1}^{m}g_i(x_{h,2}(t_k))h_i(x_{h,2}(t_k)). \tag{6.66}$$

From Eq. 6.11 we have:

$$-\alpha_3\left(\|x_{h,2}(t_k)\|\right) \leq -\alpha_3\left(\alpha_2^{-1}(\rho_s)\right) \tag{6.67}$$

for all $x_{h,2}(t_k) \in \Omega_{\rho_n}/\Omega_{\rho_s}$. From the Lipschitz properties of Eqs. 6.21–6.24 and the fact that $h_i(\cdot) \in U_i$, we obtain the following inequality:

$$\dot{V}(x_h(t)) \leq -\alpha_3\left(\alpha_2^{-1}(\rho_s)\right) + \left(L_x' + \sum_{i=1}^{m}u_i^{\max}L_{u_i}'\right)\|x_h(t)-x_{h,2}(t_k)\|$$

$$+ \sum_{i=1}^{m}C_{g_i}L_{h_i}\|\check{x}_h(t_k)-x_{h,2}(t_k)\|. \tag{6.68}$$

Using the triangular inequality, we obtain:

$$\|x_h(t)-x_{h,2}(t_k)\| \leq \|x_h(t)-x_{h,2}(t)\| + \|x_{h,2}(t)-x_{h,2}(t_k)\|, \quad \forall t \in [t_k, t_{k+1}). \tag{6.69}$$

Taking into account Eq. 6.15 and the continuity of $x_{h,2}(t)$, the following bound can be written for all $t \in [t_k, t_{k+1})$:

$$\|x_{h,2}(t)-x_{h,2}(t_k)\| \leq M\Delta. \tag{6.70}$$

Using that $\|x_h(t)-x_{h,2}(t)\| \leq f_E(0,0,\theta_x,\Delta)$ and $\|\check{x}_h(t_k)-x_{h,2}(t_k)\| \leq \theta_x$, we obtain from Eq. 6.68 the following bound on the time derivative of the Lyapunov function for $t \in [t_k, t_{k+1})$, for all initial states $x_h(t_k) \in \Omega_{\rho_n}/\Omega_{\rho_s}$:

$$\dot{V}\left(x_h(t)\right) \leq -\alpha_3\left(\alpha_2^{-1}(\rho_s)\right) + \left(L_x' + \sum_{i=1}^{m} u_i^{\max} L_{u_i}'\right)\left(f_E(0, 0, \theta_x, \Delta) + M\Delta\right)$$

$$+ \theta_x \sum_{i=1}^{m} C_{g_i} L_{h_i}. \tag{6.71}$$

If the condition of Eq. 6.58 is satisfied, then $\dot{V}(x_h(t)) \leq -\varepsilon_s/\Delta$. Integrating this bound on $t \in [t_k, t_{k+1})$, we obtain that the inequalities of Eqs. 6.59–6.60 hold. Applying this result recursively, if $x_h(t_0) \in \Omega_{\rho_n}/\Omega_{\rho_s}$, then there exists $k^* > 0$ such that $x_h(t_{k^*}) \in \Omega_{\rho_s}$, $x_h(t_k) \in \Omega_{\rho_n}/\Omega_{\rho_s}$, $\forall k \leq k^*$ and $V(x_h(t_k)) \leq V(x_h(t_0)) - k\varepsilon_s$. Once the state converges to $\Omega_{\rho_s} \subset \Omega_{\rho_{\min}}$ (or if it starts there) it remains inside $\Omega_{\rho_{\min}}$ for all times. This statement holds because from the definition of $\rho_{\min}$, if $x_h(t_k) \in \Omega_{\rho_s}$, then $x_h(t_{k+1}) \in \Omega_{\rho_{\min}}$. It follows that the condition of Eq. 6.62 holds, and thus, $x_h(t)$ is ultimately bounded in $\Omega_{\rho_{\min}}$. The bound on the evolution of the state between sampling times follows from Eq. 6.58.   □

In Theorem 6.1 below, we provide sufficient conditions under which the DMPC of Eqs. 6.28–6.35 and 6.39–6.46 guarantees that the state of the closed-loop system is ultimately bounded in a region that contains the origin. To simplify the proof of Theorem 6.1, we define new functions $f_H(\tau)$ and $f_{X2}(\tau)$ based on f_E and $f_{X,i}$ ($i = 1, \ldots, m$) (see Proposition 5.1), respectively, as follows:

$$f_H(\tau) = \sum_{i=2}^{m} \left(\frac{1}{m} L_x' + M_{g_i} L_{h_i} + u_i^{\max} L_{u_i}'\right)$$

$$\times \left(\frac{1}{L_1} f_E\left(\theta_x^i + \theta_x^1, 0, 0, \tau\right) - \frac{L_2\tau + \theta_x^i + \theta_x^1}{L_1}\right) \tag{6.72}$$

$$f_{X2}(\tau) = \left(\frac{1}{m} L_x' + L_{u_1} u_1^{\max}\right)\left(\frac{1}{C_{1,1}} f_{X,1}(0, \tau) - \frac{C_{2,1}}{C_{1,1}}\tau\right)$$

$$+ \sum_{i=2}^{m} \left(\frac{1}{m} L_x' + L_{u_i}' u_i^{\max}\right)$$

$$\times \left(\frac{1}{C_{1,i}} f_{X,i}\left(\theta_x^i + \theta_x^1, \tau\right) - \frac{C_{2,i}}{C_{1,i}}\tau - \frac{\theta_x^i + \theta_x^1}{C_{1,i}}\right). \tag{6.73}$$

It is easy to verify that $f_H(\tau)$ and $f_{X2}(\tau)$ are strictly increasing and convex functions of their arguments.

Theorem 6.1 *Consider the system of Eq. 6.3 in closed-loop with the DMPC design of Eqs. 6.28–6.35 and 6.39–6.46 based on the controller $h(x)$ that satisfies the conditions of Eqs. 6.11–6.14 with class $\mathscr{K}$ functions $\alpha_i(\cdot)$, $i = 1, 2, 3, 4$. If there exist*

$\Delta > 0$, $\varepsilon_s > 0$, $\theta_x > 0$, $\rho > \rho_n > \rho_{\min} > 0$, $\rho > \rho_n > \rho_s > 0$ and $N \geq T \geq 1$ that satisfy the conditions of Eqs. 6.58 and the following inequality:

$$-T\varepsilon_s + f_{X2}(T\Delta) + f_V\big(f_W(\theta_x, T\Delta)\big) + f_V\big(f_W(\theta_x, 0)\big) + f_H(T\Delta)$$
$$+ \sum_{i=1}^{m} C_{g,i}\,\Delta u_i (T-1)\Delta < 0 \tag{6.74}$$

with f_V defined in Eq. 2.49, f_W defined in Eq. 5.119, f_H defined in Eq. 6.72 and f_{X2} defined in Eq. 6.73, and if the initial state of the closed-loop system $x(t_0) \in \Omega_{\rho_n}$, then $x(t)$ is ultimately bounded in $\Omega_{\rho_b} \subseteq \Omega_{\rho_n}$ where:

$$\rho_b = \rho_{\min} + f_{X2}(T\Delta) + f_V\big(f_W(\theta_x, T\Delta)\big) + f_V\big(f_W(\theta_x, 0)\big) + f_H(T\Delta)$$
$$+ \sum_{i=1}^{m} C_{g,i}\,\Delta u_i (T-1)\Delta \tag{6.75}$$

with $\rho_{\min}$ defined in Eq. 6.61.

Proof We first consider two consecutive time instants in which both fast and slowly sampled states are available and we have full system state measurements which are affected by noise: t_k and t_{k+T} ($k = 0, T, 2T, \ldots$). We will prove that the Lyapunov function of the system is decreasing from t_k to t_{k+T}. In the following, we denote the trajectory of the nominal system of Eq. 6.3 under the DMPC of Eqs. 6.28–6.35 and 6.39–6.46 starting from $\check{x}^1(t_k)$ (which is the state received by LMPC 1 at t_k) as $\tilde{x}$, and we also denote the predicted nominal system trajectory in the evaluation of the LMPC of Eqs. 6.28–6.34 at the final iteration as $\tilde{x}^j$ with $j = 1, \ldots, m$. It should be mentioned that the initial condition for the nominal sampled trajectory $\tilde{x}$ under the implementation of u_i^* can be $\tilde{x}(t_k) = x_h^i(t|t_k)$ for any $i = 1, \ldots, m$. Without loss of generality, we assume that $\tilde{x}(t_k) = \check{x}^1(t_k) = x_h^1(t|t_k)$; use of any $i = 2, \ldots, m$ in $\tilde{x}(t_k) = x_h^i(t|t_k)$ would simply require an appropriate modification in the definitions of $f_{X2}(\cdot)$ and $f_H(\cdot)$.

The derivative of the Lyapunov function of the nominal system of Eq. 6.3 under the multirate DMPC of Eqs. 6.28–6.35 and 6.39–6.46 from t_k to t_{k+T} can be expressed as follows:

$$\dot{V}\big(\tilde{x}(t)\big) = \frac{\partial V(\tilde{x}(t))}{\partial x}\left(f\big(\tilde{x}(t)\big) + \sum_{i=1}^{m} g_i\big(\tilde{x}(t)\big)u_i^*(t) \right), \tag{6.76}$$

where $\tilde{x}(t_k) = \check{x}^1(t_k) = x_h^1(t_k|t_k)$ and $u_i^*(t)$ is the actual input applied to the system and defined as follows:

$$u_i^*(t) = \begin{cases} u_i^{*,f}(t|t_k), & t \in [t_k, t_{k+1}), \\ u_i^{*,l}(t|t_l), & t \in [t_l, t_{l+1}), l = k+1, \ldots, k+T-1. \end{cases} \tag{6.77}$$

Combining Eq. 6.76 and the inequality constraints of Eq. 6.34 ($i = 1, \ldots, m$), and adding and subtracting the term $\frac{\partial V(x_h^1(t|t_k))}{\partial x}(f(x_h^1(t|t_k)) + \sum_{i=1}^{m} g_i(x_h^1(t|t_k)) \times u_{h,i}^1(t|t_k))$ to and from the right-hand side of the resulting inequality, we can obtain the following inequality for all $t \in [t_k, t_{k+T})$:

$$
\begin{aligned}
\dot{V}\big(\tilde{x}(t)\big) \leq{} & \frac{\partial V(x_h^1(t|t_k))}{\partial x}\left(f\big(x_h^1(t|t_k)\big) + \sum_{i=1}^{m} g_i\big(x_h^1(t|t_k)\big)u_{h,i}^1(t|t_k)\right) \\
& + \frac{\partial V(x_h^1(t|t_k))}{\partial x}\left(\frac{1}{m}f\big(x_h^1(t|t_k)\big) + g_1\big(x_h^1(t|t_k)\big)u_{h,1}^1(t|t_k)\right) \\
& - \frac{\partial V(x_h^1(t|t_k))}{\partial x}\left(\frac{1}{m}f\big(x_h^1(t|t_k)\big) + g_1\big(x_h^1(t|t_k)\big)u_{h,1}^1(t|t_k)\right) \\
& + \cdots \\
& + \frac{\partial V(x_h^m(t|t_k))}{\partial x}\left(\frac{1}{m}f\big(x_h^m(t|t_k)\big) + g_m\big(x_h^m(t|t_k)\big)u_{h,m}^m(t|t_k)\right) \\
& - \frac{\partial V(x_h^1(t|t_k))}{\partial x}\left(\frac{1}{m}f\big(x_h^1(t|t_k)\big) + g_m\big(x_h^1(t|t_k)\big)u_{h,m}^1(t|t_k)\right) \\
& + \frac{\partial V(\tilde{x}(t))}{\partial x}\left(\frac{1}{m}f\big(\tilde{x}(t)\big) + g_1(\tilde{x})u_1^{*,f}(t|t_k)\right) \\
& - \frac{\partial V(\tilde{x}^1(t))}{\partial x}\left(\frac{1}{m}f\big(\tilde{x}^1(t)\big) + g_1\big(\tilde{x}^1(t)\big)u_1^{*,f}(t|t_k)\right) \\
& + \cdots \\
& + \frac{\partial V(\tilde{x}(t))}{\partial x}\left(\frac{1}{m}f\big(\tilde{x}(t)\big) + g_m(\tilde{x})u_m^{*,f}(t|t_k)\right) \\
& - \frac{\partial V(\tilde{x}^m(t))}{\partial x}\left(\frac{1}{m}f\big(\tilde{x}^m(t)\big) + g_m\big(\tilde{x}^m(t)\big)u_m^{*,f}(t|t_k)\right) \\
& + \sum_{i=1}^{m} \frac{\partial V(\tilde{x}(t))}{\partial x}g_i\big(\tilde{x}(t)\big)\big(u_i^*(t) - u_i^{*,f}(t|t_k)\big).
\end{aligned}
\tag{6.78}
$$

Using the locally Lipschitz properties of Eqs. 6.21–6.24, the following inequality can be obtained for $t \in [t_k, t_{k+T})$ from the inequality of Eq. 6.78:

$$
\begin{aligned}
\dot{V}\big(\tilde{x}(t)\big) \leq{} & \dot{V}\big(x_h^1(t|t_k)\big) \\
& + \left(\frac{1}{m}L_x' + L_{u_1}' u_1^{*,f}(t|t_k)\right)\big\|\tilde{x}(t) - \tilde{x}^1(t)\big\| + \cdots \\
& + \left(\frac{1}{m}L_x' + L_{u_m}' u_m^{*,f}(t|t_k)\right)\big\|\tilde{x}(t) - \tilde{x}^m(t)\big\|
\end{aligned}
$$

$$+ \left(\frac{1}{m} L'_x + M_{g2} L_{h2} + u_2^{\max} L'_{u_2} \right) \left\| x_h^2(t|t_k) - x_h^1(t|t_k) \right\| + \cdots$$

$$+ \left(\frac{1}{m} L'_x + M_{gm} L_{hm} + u_m^{\max} L'_{u_m} \right) \left\| x_h^m(t|t_k) - x_h^1(t|t_k) \right\|$$

$$+ \sum_{i=1}^{m} C_{g_i} \left(u_i^*(t) - u_i^{*,f}(t|t_k) \right). \tag{6.79}$$

Applying Propositions 5.1 and 6.1 to the inequality of Eq. 6.79, we have:

$$\dot{V}\left(\tilde{x}(t)\right) \le \dot{V}\left(x_h^1(t|t_k)\right)$$

$$+ \left(\frac{1}{m} L'_x + L'_{u_1} u_1^{*,f}(t|t_k) \right) f_{X,1}(0, t)$$

$$+ \left(\frac{1}{m} L'_x + L'_{u_2} u_2^{*,f}(t|t_k) \right) f_{X,2}\left(\theta_x^1 + \theta_x^2, t\right) + \cdots$$

$$+ \left(\frac{1}{m} L'_x + L'_{u_m} u_m^{*,f}(t|t_k) \right) f_{X,m}\left(\theta_x^1 + \theta_x^m, t\right)$$

$$+ \left(\frac{1}{m} L'_x + M_{g2} L_{h2} + u_2^{\max} L'_{u_2} \right) f_E\left(\theta_x^1 + \theta_x^2, 0, 0, t\right) + \cdots$$

$$+ \left(\frac{1}{m} L'_x + M_{gm} L_{hm} + u_m^{\max} L'_{u_m} \right) f_E\left(\theta_x^1 + \theta_x^m, 0, 0, t\right) + \cdots$$

$$+ \sum_{i=1}^{m} C_{g_i} \left(u_i^*(t) - u_i^{*,f}(t|t_k) \right). \tag{6.80}$$

Integrating the inequality of Eq. 6.80 from $t = t_k$ to $t = t_{k+T}$ and taking into account that $\tilde{x}(t_k) = x_h^1(t_k|t_k)$, the constraints of Eqs. 6.31 and 6.43 and the definitions of $f_{X2}(\cdot)$, $f_H(\cdot)$ and $u^*(t)$, the following inequality can be obtained:

$$V\left(\tilde{x}(t_{k+T})\right) \le V\left(x_h^1(t_{k+T}|t_k)\right) + f_{X2}(T\Delta) + f_H(T\Delta) + \sum_{i=1}^{m} C_{g,i} \Delta u_i (T-1)\Delta. \tag{6.81}$$

Using the above inequality and since $V(x_h^1(t_{k+T}|t_k)) \le \max\{V(x_h^1(t_k|t_k)) - T\varepsilon_s, \rho_{\min}\}$ from Corollary 6.1, $\tilde{x}(t_k) = x_h^1(t_k|t_k)$ and $\|V(\tilde{x}(t_k)) - V(x(t_k))\| \le f_V(f_W(\theta_x, 0))$ and $\|V(\tilde{x}(t_{k+T})) - V(x(t_{k+T}))\| \le f_V(f_W(\theta_x, T\Delta))$ from Corollary 5.3 and Proposition 2.3, we can obtain the following inequality:

$$V\left(x(t_{k+T})\right) \le \max\left\{V\left(x(t_k)\right) - T\varepsilon_s, \rho_{\min}\right\}$$

$$+ f_{X2}(T\Delta) + f_H(T\Delta) + f_V\left(f_W(\theta_x, T\Delta)\right)$$

$$+ f_V\left(f_W(\theta_x, 0)\right) + \sum_{i=1}^{m} C_{g,i} \Delta u_i (T-1)\Delta. \tag{6.82}$$

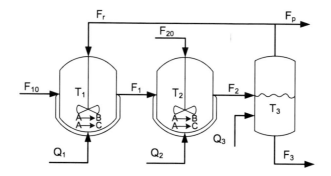

Fig. 6.2 Reactor–separator with recycle process network with reactions $A \rightarrow B$ and $A \rightarrow C$

If there exist $\Delta > 0$, $\varepsilon_s > 0$, $\theta_x > 0$, $\rho > \rho_n > \rho_{\min} > 0$, $\rho > \rho_n > \rho_s > 0$ and $N \geq T \geq 1$ that satisfy the conditions of Eqs. 6.58 and Eq. 6.74, then there exists $\varepsilon_w > 0$ such that the following inequality holds:

$$V\left(x(t_{k+T})\right) \leq \max\left\{V\left(x(t_k)\right) - \varepsilon_w, \rho_b\right\}, \qquad (6.83)$$

which implies that if $x(t_k) \in \Omega_{\rho_n}/\Omega_{\rho_b}$, then $V(x(t_{k+T})) < V(x(t_k))$, and if $x(t_k) \in \Omega_{\rho_b}$, then $V(x(t_{k+T})) \leq \rho_b$.

Because the upper bound on the difference between the Lyapunov function of the actual trajectory x and the nominal trajectory $\tilde{x}$ (the term $f_{X2}(T\Delta) + f_H(T\Delta) + f_V(f_W(\theta_x, T\Delta)) + f_V(f_W(\theta_x, 0)) + \sum_{i=1}^{m} C_{g,i}\Delta u_i(T-1)\Delta)$ is a strictly increasing function of T, the inequality of Eq. 6.83 also implies that:

$$V\left(x(t)\right) \leq \max\left\{V\left(x(t_k)\right) - \varepsilon_w, \rho_b\right\}, \quad \forall t \in [t_k, t_{k+T}). \qquad (6.84)$$

Using the inequality of Eq. 6.84 recursively, it can be proved that if $x(t_0) \in \Omega_{\rho_n}$, then the closed-loop trajectories of the system of Eq. 6.3 under the multirate DMPC design stay in Ω_{ρ_n} for all times (i.e., $x(t) \in \Omega_{\rho_n}$ for all t). Moreover, if $x(t_0) \in \Omega_{\rho_n}$, the closed-loop trajectories of the system of Eq. 6.3 under the multirate DMPC design satisfy:

$$\limsup_{t \to \infty} V\left(x(t)\right) \leq \rho_b. \qquad (6.85)$$

This proves the stability result stated in Theorem 6.1.   □

6.6 Application to a Reactor–Separator Process

Consider the three vessel, reactor–separator process with different reactions taking place in the vessels as shown in Fig. 6.2 [10]. A feed stream to the first CSTR contains the reactant, A, which is converted into the desired product, B. Species A can also react into an undesired side-product, C. The solvent does not react and is labeled as D. The effluent of the first CSTR along with additional fresh feed makes up the inlet to the second CSTR. The reactions $A \rightarrow B$ and $A \rightarrow C$ (referred to as 1 and 2, respectively) take place in the two CSTRs in series before the effluent from CSTR 2 is fed to a flash tank. The overhead vapor from the flash tank is condensed and recycled to the first CSTR, and the bottom product stream is removed. All three vessels are assumed to have static holdup. The dynamic equations describing the behavior of the system, obtained through material and energy balances under standard modeling assumptions, are given below:

$$\frac{dT_1}{dt} = \frac{F_{10}}{V_1}(T_{10} - T_1) + \frac{F_r}{V_1}(T_3 - T_1) + \frac{-\Delta H_1}{\rho C_p} k_1 e^{\frac{-E_1}{RT_1}} C_{A1}$$

$$+ \frac{-\Delta H_2}{\rho C_p} k_2 e^{\frac{-E_2}{RT_1}} C_{A1} + \frac{Q_1}{\rho C_p V_1}, \tag{6.86}$$

$$\frac{dC_{A1}}{dt} = \frac{F_{10}}{V_1}(C_{A10} - C_{A1}) + \frac{F_r}{V_1}(C_{Ar} - C_{A1})$$

$$- k_1 e^{\frac{-E_1}{RT_1}} C_{A1} - k_2 e^{\frac{-E_2}{RT_1}} C_{A1}, \tag{6.87}$$

$$\frac{dC_{B1}}{dt} = \frac{-F_{10}}{V_1} C_{B1} + \frac{F_r}{V_1}(C_{Br} - C_{B1}) + k_1 e^{\frac{-E_1}{RT_1}} C_{A1}, \tag{6.88}$$

$$\frac{dC_{C1}}{dt} = \frac{-F_{10}}{V_1} C_{C1} + \frac{F_r}{V_1}(C_{Cr} - C_{C1}) + k_2 e^{\frac{-E_2}{RT_1}} C_{A1}, \tag{6.89}$$

$$\frac{dT_2}{dt} = \frac{F_1}{V_2}(T_1 - T_2) + \frac{F_{20}}{V_2}(T_{20} - T_2) + \frac{-\Delta H_1}{\rho C_p} k_1 e^{\frac{-E_1}{RT_2}} C_{A2}$$

$$+ \frac{-\Delta H_2}{\rho C_p} k_2 e^{\frac{-E_2}{RT_2}} C_{A2} + \frac{Q_2}{\rho C_p V_2}, \tag{6.90}$$

$$\frac{dC_{A2}}{dt} = \frac{F_1}{V_2}(C_{A1} - C_{A2}) + \frac{F_{20}}{V_2}(C_{A20} - C_{A2})$$

$$- k_1 e^{\frac{-E_1}{RT_2}} C_{A2} - k_2 e^{\frac{-E_2}{RT_2}} C_{A2}, \tag{6.91}$$

$$\frac{dC_{B2}}{dt} = \frac{F_1}{V_2}(C_{B1} - C_{B2}) - \frac{F_{20}}{V_2} C_{B2} + k_1 e^{\frac{-E_1}{RT_2}} C_{A2}, \tag{6.92}$$

$$\frac{dC_{C2}}{dt} = \frac{F_1}{V_2}(C_{C1} - C_{C2}) - \frac{F_{20}}{V_2} C_{C2} + k_2 e^{\frac{-E_2}{RT_2}} C_{A2}, \tag{6.93}$$

$$\frac{dT_3}{dt} = \frac{F_2}{V_3}(T_2 - T_3) - \frac{H_{vap} F_r}{\rho C_p V_3} + \frac{Q_3}{\rho C_p V_3}, \tag{6.94}$$

$$\frac{dC_{A3}}{dt} = \frac{F_2}{V_3}(C_{A2} - C_{A3}) - \frac{F_r}{V_3}(C_{Ar} - C_{A3}), \tag{6.95}$$

$$\frac{dC_{B3}}{dt} = \frac{F_2}{V_3}(C_{B2} - C_{B3}) - \frac{F_r}{V_3}(C_{Br} - C_{B3}), \tag{6.96}$$

$$\frac{dC_{C3}}{dt} = \frac{F_2}{V_3}(C_{C2} - C_{C3}) - \frac{F_r}{V_3}(C_{Cr} - C_{C3}).$$ (6.97)

Each of the tanks has an external heat input/removal actuator. The model of the flash tank separator is derived under the assumption that the relative volatility for each of the species remains constant within the operating temperature range of the flash tank. This assumption allows calculating the mass fractions in the overhead based upon the mass fractions in the liquid portion of the vessel. It has also been assumed that there is a negligible amount of reaction taking place in the separator. The following algebraic equations model the composition of the overhead stream relative to the composition of the liquid holdup in the flash tank:

$$C_{Ar} = \frac{\alpha_A C_{A3}}{K},$$ (6.98)

$$C_{Br} = \frac{\alpha_B C_{B3}}{K},$$ (6.99)

$$C_{Cr} = \frac{\alpha_C C_{C3}}{K},$$ (6.100)

where

$$K = \alpha_A C_{A3}\frac{MW_A}{\rho} + \alpha_B C_{B3}\frac{MW_B}{\rho} + \alpha_C C_{C3}\frac{MW_C}{\rho} + \alpha_D x_D \rho,$$ (6.101)

and x_D is the mass fraction of the solvent in the flash tank liquid holdup and is found from a mass balance. The definitions for the variables used in Eqs. 6.86–6.100 be found in Table 6.1, with the parameter values given in Table 6.2.

The system of Eqs. 6.86–6.100 is numerically simulated using a standard Euler integration method. Process noise was added to the right-hand side of each equation in the differential equations to simulate disturbances/model uncertainty and it is generated as autocorrelated noise of the form $w_k = \phi w_{k-1} + \xi_k$ where $k = 0, 1, \ldots$ is the discrete time step of 0.001 h, ξ_k is generated by a normally distributed random variable with standard deviation σ_p, and ϕ is the autocorrelation factor and w_k is bounded by θ_p, that is $\|w_k\| \leq \theta_p$. Table 6.3 contains the parameters used in generating the process noise.

This process is divided into three subsystems corresponding to the first CSTR, the second CSTR and the separator, respectively. For the three subsystems, we will refer to them as subsystem 1, subsystem 2 and subsystem 3, respectively. The state of subsystem 1 is defined as the deviations of the temperature and species concentrations in the first CSTR from their desired steady-state; that is, $x_1^T = [x_{f,1}^T, x_{s,1}^T]$ where $x_{f,1} = T_1 - T_{1s}$ and $x_{s,1}^T = [C_{A1} - C_{A1s} \ C_{B1} - C_{B1s} \ C_{C1} - C_{Cs}]$ denote fast sampled and slowly sampled measurements of subsystem 1, respectively. Due to the simplicity of temperature measurement at each sampling time, we denote the temperature as the fast sampled measurement of each subsystem. The states of subsystems 2 and 3 are defined similarly; they are $x_2^T = [T_2 - T_{2s} \ C_{A2} - C_{A2s} \ C_{B2} - C_{B2s} \ C_{C2} - C_{C2s}]$ and $x_3^T = [T_3 - T_{3s} \ C_{A3} - C_{A3s} \ C_{B3} - C_{B3s} \ C_{C3} - C_{C3s}]$. Accordingly, the state of

Table 6.1 Process variables of the reactor–separator process of Eqs. 6.86–6.100

C_{A1}, C_{A2}, C_{A3}	Concentrations of A in vessels 1, 2, 3
C_{B1}, C_{B2}, C_{B3}	Concentrations of B in vessels 1, 2, 3
C_{C1}, C_{C2}, C_{C3}	Concentrations of C in vessels 1, 2, 3
C_{Ar}, C_{Br}, C_{Cr}	Concentrations of A, B, C in the recycle
T_1, T_2, T_3	Temperatures in vessels 1, 2, 3
T_{10}, T_{20}	Feed stream temperatures to vessels 1, 2
F_1, F_2, F_3	Effluent flow rates from vessels 1, 2, 3
F_{10}, F_{20}	Feed stream flow rates to vessels 1, 2
C_{A10}, C_{A20}	Concentrations of A in the feed stream to vessels 1, 2
F_r	Recycle flow rate
V_1, V_2, V_3	Volumes of vessels 1, 2, 3
E_1, E_2	Activation energy for reactions 1, 2
k_1, k_2	Pre-exponential values for reactions 1, 2
$\Delta H_1, \Delta H_2$	Heats of reaction for reactions 1, 2
H_{vap}	Heat of vaporization
$\alpha_A, \alpha_B, \alpha_C, \alpha_D$	Relative volatilities of A, B, C, D
MW_A, MW_B, MW_C	Molecular weights of $A, B,$ and C
Q_1, Q_2, Q_3	Heat inputs into vessels 1, 2, 3
C_p, R, ρ	Heat capacity, gas constant and solution density

Table 6.2 Parameter values of the reactor–separator process of Eqs. 6.86–6.100

T_{10}	300 [k]	T_{20}	300 [K]
F_{10}	5 [m³/h]	F_{20}	5 [m³/h]
F_r	1.9 [m³/h]	C_{A10}	4 [kmol/m³]
C_{A20}	3 [kmol/m³]	V_1	1.0 [m³]
V_2	0.5 [m³]	V_3	1.0 [m³]
E_1	5×10^4 [KJ/kmol]	E_2	5.5×10^4 [KJ/kmol]
k_1	3×10^6 [h⁻¹]	k_2	3×10^6 [h⁻¹]
ΔH_1	-5×10^4 [KJ/kmol]	ΔH_2	-5.3×10^4 [KJ/kmol]
H_{vap}	5 [KJ/kmol]	C_p	0.231 [KJ/kg K]
R	8.314 [KJ/kmol K]	ρ	1000 [kg/m³]
α_A	2	α_B	1
α_C	1.5	α_D	3
MW_A	50 [kg/kmol]	MW_B	50 [kg/kmol]
MW_C	50 [kg/kmol]		

the whole process is defined as a combination of the states of the three subsystems; that is, $x^T = [x_1^T \; x_2^T \; x_3^T]$.

Table 6.3 Noise parameters of the reactor–separator process of Eqs. 6.86–6.100

	σ_p	ϕ	θ_p
C_{A1}	0.1	0.7	0.09
C_{B1}	0.02	0.7	0.01
C_{C1}	0.02	0.7	0.01
T_1	10	0.7	1.17
C_{A2}	0.1	0.7	0.09
C_{B2}	0.1	0.7	0.03
C_{C2}	0.1	0.7	0.01
T_2	10	0.7	1.35
C_{A3}	0.1	0.7	0.09
C_{B3}	0.1	0.7	0.02
C_{C3}	0.02	0.7	0.01
T_3	10	0.7	1.35

Table 6.4 Steady-state values for x_s of the reactor–separator process of Eqs. 6.86–6.100

C_{A1s}	3.31 [kmol/m^3]	C_{A2s}	2.75 [kmol/m^3]
C_{A3s}	2.88 [kmol/m^3]	C_{B1s}	0.17 [kmol/m^3]
C_{B2s}	0.45 [kmol/m^3]	C_{B3s}	0.50 [kmol/m^3]
C_{C1s}	0.04 [kmol/m^3]	C_{C2s}	0.11 [kmol/m^3]
C_{C3s}	0.12 [kmol/m^3]	T_{1s}	369.53 [K]
T_{2s}	435.25 [K]	T_{3s}	435.25 [K]

Table 6.5 Steady-state values for Q_{1s}, Q_{2s} and Q_{3s} of the reactor–separator process of Eqs. 6.86–6.100

Q_{1s}	0 [KJ/h]	Q_{2s}	0 [KJ/h]
Q_{3s}	0 [KJ/h]		

The external heat input associated with each vessel is the control input associated with each subsystem, that is, $u_1 = Q_1 - Q_{1s}$, $u_2 = Q_2 - Q_{2s}$ and $u_3 = Q_3 - Q_{3s}$. The process has one unstable and two stable steady states. The control objective is to regulate the process at the unstable steady-state x_s shown in Table 6.4 with corresponding steady-state inputs as shown in Table 6.5. The inputs are subject to constraints as follows: $|u_1| \leq 5 \times 10^4$ KJ/h, $|u_2| \leq 1.5 \times 10^5$ KJ/h, and $|u_3| \leq 2 \times 10^5$ KJ/h. Three distributed LMPCs (controller 1, controller 2 and controller 3) will be designed to manipulate each one of the three inputs in the three subsystems, respectively. The process model of Eqs. 6.86–6.100 belongs to the following class of nonlinear systems:

$$\dot{x}(t) = f(x(t)) + \sum_{i=1}^{3} g_i(x(t))u_i(t) + w(x(t)),$$

where the explicit expressions of f, g_i ($i = 1, 2, 3$), are omitted for brevity.

We assume that $x_{f,1}, x_{f,2}, x_{f,3}$ are measured and sent to controller 1, controller 2 and controller 3, respectively, at synchronous time instants $t_l = l\Delta$, $l = 0, 1, \ldots$, with $\Delta = 0.01$ h $= 36$ s while we assume that each controller receives $x_{s,i}$ every $T = 4$ sampling times. The three subsystems exchange their states at $t_k = kT\Delta$, $k = 0, 1, \ldots$; that is, the full system state x is sent to all the controllers every $T = 4$ sampling times.

In the simulations, we consider a quadratic Lyapunov function $V(x) = x^T P x$ with:

$$P = diag\left(\begin{bmatrix} 20 \ 10^3 \ 10^3 \ 10^3 \ 20 \ 10^3 \ 10^3 \ 10^3 \ 20 \ 10^3 \ 10^3 \ 10^3 \end{bmatrix}\right). \tag{6.102}$$

We design the Lyapunov-based controller $h(x)$ following the continuous bounded control law [47] as follows:

$$h(x) = -p(x)(L_G V)^T, \tag{6.103}$$

where

$$p(x) = \begin{cases} \dfrac{L_f V + \sqrt{(L_f V)^2 + (u^{\max} \| L_G V^T \|)^4}}{\| L_G V^T \|^2 [1 + \sqrt{1 + (u^{\max} \| L_G V^T \|)^2}]}, & L_G V \neq 0, \\ 0, & L_G V = 0 \end{cases}$$

with $L_f V = \frac{\partial V}{\partial x} f(x)$ and $L_G V = \frac{\partial V}{\partial x} G(x)$ where $G = [g_1 \ g_2 \ g_3]$ being the Lie derivatives of the scalar function V with respect to the vector fields f and G, respectively.

Based on $h(x)$ and $V(x)$, we design the three LMPCs following Eqs. 6.28–6.35 and 6.39–6.46 and refer to them as LMPC 1, LMPC 2 and LMPC 3. For each LMPC, we also design a state observer following Eqs. 6.36–6.37. In the design of the LMPCs, the weighting matrices are chosen to be:

$$Q_c = diag\left(\begin{bmatrix} 20 \ 10^3 \ 10^3 \ 10^3 \ 20 \ 10^3 \ 10^3 \ 10^3 \ 20 \ 10^3 \ 10^3 \ 10^3 \end{bmatrix}\right), \tag{6.104}$$

and $R_1 = R_2 = R_3 = 10^{-6}$. The prediction horizon for the optimization problem is $N = 5$ with a time step of $\Delta = 0.01$ h. In the simulations, we put a maximum iteration number $c_{\max}$ on the DMPC evaluation and the maximum iteration number is chosen to be $c_{\max} = 2$. Also, we set Δu_i as 10% percent of $u_i^{\max}$ ($i = 1, 2, 3$). The optimization problems are solved by the open source interior point optimizer Ipopt [109]. The initial condition which is utilized to carry out simulations is as follows:

$$x(0)^T = [360.69 \ 3.19 \ 0.15 \ 0.03 \ 430.91 \ 2.76 \ 0.34 \ 0.08 \ 430.42 \ 2.79 \ 0.38 \ 0.08]. \tag{6.105}$$

We set the bound on the measurement noise to be 1% of the instantaneous value of the signal measured by sensors. The communication channel noise is generated using gaussian random variables with variances σ_n and σ_u bounded by θ_n and θ_u for state values and control inputs, respectively. These values are shown in Tables 6.6 and 6.7.

Figures 6.3 and 6.4 show the temperature, concentration and input trajectories of the process under the multirate DMPC design of Eqs. 6.28–6.35 and 6.39–6.46 in

Table 6.6 Communication noise parameters for states of the reactor–separator process of Eqs. 6.86–6.100

	σ_n	θ_n		σ_n	θ_n
C_{A1}	1	0.033	C_{A2}	1	0.027
C_{A3}	1	0.028	C_{B1}	1	0.001
C_{B2}	1	0.004	C_{B3}	1	0.005
C_{C1}	1	0.001	C_{C2}	1	0.001
C_{C3}	1	0.001	T_1	10	3.695
T_2	10	4.352	T_3	10	4.352

Table 6.7 Communication noise parameters for control inputs of the reactor–separator process of Eqs. 6.86–6.100

	σ_u	θ_u
u_1	10	7.39
u_2	30	22.17
u_3	40	29.56

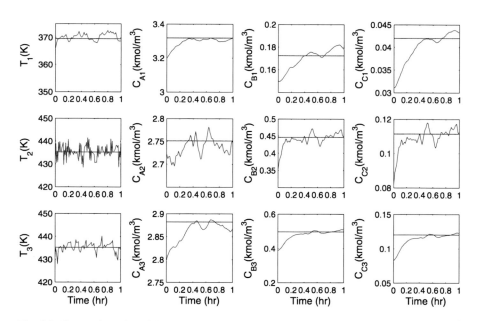

Fig. 6.3 State trajectories of the reactor–separator process of Eqs. 6.86–6.100 under the multirate DMPC of Eqs. 6.28–6.35 and 6.39–6.46

the presence of communication and measurement noise. As it can be seen from these figures, the multirate DMPC system can steer the system state to a neighborhood of the desired steady-state.

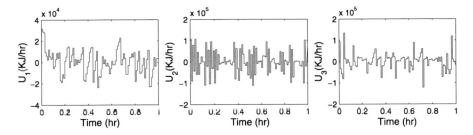

Fig. 6.4 Manipulated input trajectories of the reactor–separator process of Eqs. 6.86–6.100 under the multirate DMPC of Eqs. 6.28–6.35 and 6.39–6.46

We also carried out a set of simulations to demonstrate the optimality of the closed-loop performance of the multirate DMPC compared with different control schemes. Specifically, we compared the multirate DMPC with five different control schemes from a performance point of view for the case in which there is no communication and measurement noise. The six control schemes considered are as follows: (1) the multirate DMPC of Eqs. 6.28–6.35 and 6.39–6.46; (2) a DMPC design with LMPCs formulated as in Eqs. 6.28–6.34 which are only evaluated at time instants in which full system states are available and the inputs are implemented in open-loop fashion between two full system state measurements (in this case, the additional fast sampled measurements are not used to improve the closed-loop performance); (3) multirate DMPC design but without communication between the distributed controllers and each controller estimating the full system states and the actions of the other controllers based on the process model and $h(x)$ (in this case, a distributed LMPC in the DMPC design takes advantage of both fast and slowly sampled measurements of its own local subsystem but does not receive any input or state information from the other subsystems); (4) the DMPC design as in (2) but without communication between the distributed controllers and each controller estimating the full system states and actions of the other controllers based on the process model and $h(x)$; (5) $h(x)$ applied in sample-and-hold; (6) a centralized LMPC introduced in [72]. We perform these simulations under different initial conditions and different process noise/disturbances. To carry out this comparison, we have computed the total cost of each simulation based on the index of the following form:

$$J = \int_{t_0}^{t_M} \left[\|x(\tau)\|_{Q_c} + \sum_{j=1}^{3} \|u_j(\tau)\|_{R_{cj}} \right] d\tau, \qquad (6.106)$$

where $t_0 = 0$ is the initial time of the simulations and $t_M = 1$ h is the end of the simulations. Table 6.8 shows the total cost computed for 10 different closed-loop simulations under the six different control schemes. From Table 6.8, we see that the centralized LMPC gives the best performance and the multirate DMPC design gives the second best performance in all the simulations. Also, Table 6.8 demonstrates that when there is communication between controllers or there is MPC implementation when there is only partial state information in each controller (fast sampled state),

Table 6.8 Total performance costs along the closed-loop trajectories of the reactor–separator process of Eqs. 6.86–6.100 under different control schemes

sim.	(1)	(2)	(3)	(4)	(5)	(6)
1	43963	633589	72200	812903	1116578	27057
2	21512	606628	28079	743874	1095819	7370
3	23041	604148	27407	706319	1084445	15112
4	24681	613289	30211	720131	1104045	8838
5	31440	618649	36290	723598	1106508	18654
6	21775	654268	25950	859380	1079984	15287
7	28553	667143	34209	879852	1109976	13168
8	28974	659250	34565	865643	1109363	13424
9	28228	672756	33949	891549	1110884	12991
10	23929	668499	29688	887300	1106623	11903

the closed-loop performance is improved. Note that the Lyapunov-based controller is a feasible solution to the DMPC problem; however, the DMPC solution can substantially improve closed-loop performance while it inherits closed-loop stability from the Lyapunov-based controller. All of the DMPC designs yield improvement in performance compared to the Lyapunov-based controller.

6.7 Conclusions

In this chapter, we designed a DMPC system using multirate sampling for large-scale nonlinear uncertain systems composed of several coupled subsystems. Specifically, we considered that the states of each local subsystem can be divided into fast sampled states (which are available every sampling time) and slowly sampled states (which are available every several sampling times). The distributed controllers communicate over a shared communication network in an iterative manner at time instants in which full system state measurements (both fast and slow) are available and the controllers communicate, to guarantee closed-loop stability. When local subsystem fast sampled state information is available, the distributed controllers operate in a decentralized fashion to improve closed-loop performance. In the multirate control architecture, the controllers were designed via LMPC techniques taking into account bounded measurement and communication noise and process disturbances. Sufficient conditions under which the state of the closed-loop system is ultimately bounded in an invariant region containing the origin were derived. Finally, the applicability and performance of the proposed DMPC scheme were demonstrated through a nonlinear chemical process network example.

Chapter 7
Conclusions

This book presented approaches to networked and distributed predictive control of nonlinear process systems via model predictive control and Lyapunov-based control techniques. Following an introduction to the motivation and objectives of this book, Lyapunov-based predictive control methods for nonlinear systems which provide an explicit characterization for the closed-loop stability region and account for the effect of asynchronous feedback and time-varying measurement delays were first developed. Then, a two-tier framework for the design of networked predictive control systems for nonlinear processes that naturally augment dedicated control systems with networked control systems was presented. Subsequently, distributed predictive control methods for large-scale nonlinear process networks taking into account asynchronous measurements and time-varying delays as well as different sampling rates of measurements were presented. Throughout the book, the effectiveness and performance of the control approaches were illustrated via applications to nonlinear process networks and wind-solar energy generation systems.

Specifically, in Chap. 2, two LMPC designs for nonlinear systems subject to data losses and time-varying measurement delays were presented. In order to provide guaranteed closed-loop stability results in the presence of data losses and/or time-varying measurement delays, the constraints that define the LMPC optimization problems as well as the implementation procedures were modified to account for data losses/asynchronous measurements and time-varying measurement delays. The presented LMPCs possess an explicit characterization of the closed-loop system stability regions. Using a nonlinear CSTR example, it was demonstrated that the presented LMPC approaches are robust to data losses and measurement delays.

In Chap. 3, a two-tier networked control architecture, which naturally augments preexisting, point-to-point control systems with networked control systems, was presented. The two-tier networked control architecture is a decentralized control architecture which is able to take advantage of asynchronous and delayed measurements and additional actuation capabilities provided by real-time wired or wireless sensor and actuator networks. Using a nonlinear CSTR example and a nonlinear reactor–separator example, the two-tier control architecture was demonstrated to be more optimal compared with conventional control systems and to be more robust

P.D. Christofides et al., *Networked and Distributed Predictive Control*,
Advances in Industrial Control,
DOI 10.1007/978-0-85729-582-8_7, © Springer-Verlag London Limited 2011

compared with centralized predictive control systems. The two-tier control architecture was also applied to the problem of optimal management and operation of a standalone wind-solar energy generation system.

In Chap. 4, a DMPC design involving two controllers was presented where the preexisting LCS and the new NCS were redesigned/designed via LMPC. This DMPC design uses a hierarchical control architecture in the sense that the LCS stabilizes the closed-loop system and the NCS takes advantage of additional control inputs to improve the closed-loop performance and provide the potential of maintaining the desired closed-loop stability and performance levels in the face of new/failing actuators. The extensions of this DMPC architecture to account for asynchronous and delayed measurements were also discussed. Using a nonlinear reactor–separator example, the stability, performance and robustness of the DMPC designs were illustrated.

In Chap. 5, sequential and iterative DMPC designs for large-scale nonlinear systems in which several distinct sets of manipulated inputs are used to regulate the overall system were presented. In the sequential DMPC architecture, the distributed controllers communicate via a one-directional communication network and are evaluated in sequence; in the iterative DMPC architecture, the distributed controllers communicate via a bidirectional communication network, are evaluated in parallel and iterate to improve closed-loop performance. Sequential and iterative DMPC designs accounting for asynchronous and delayed measurements were also considered. In addition, an approach to handle communication disruptions and data losses between the distributed controllers was discussed in the framework of the hierarchical DMPC architecture of Chap. 4. Using a nonlinear catalytic alkylation of benzene process example, the DMPC designs were compared with the corresponding centralized MPC designs from stability, evaluation time, and convergence points of view.

In Chap. 6, a multirate DMPC design for large-scale nonlinear uncertain systems with fast and slowly sampled states was developed. The distributed model predictive controllers are connected through a shared communication network and cooperate in an iterative fashion at time instants in which both fast and slowly sampled measurements are available, to guarantee closed-loop stability. When only local subsystem fast sampled state information is available, the distributed controllers operate in a decentralized fashion to improve closed-loop performance. Using a reactor–separator process example, the stability property and performance of the multirate DMPC architecture was illustrated.

Future research in networked and distributed predictive process control as well as related areas includes the development of general methods for the handling of broad class of communication disruptions between distributed controllers, the design of distributed state estimation systems which provide fast and guaranteed convergence and the development of distributed plant monitoring and fault-tolerant control systems. The reader may refer to [12, 92, 95] for more discussions on the related open problems.

References

1. Allgöwer, F., & Chen, H. (1998). Nonlinear model predictive control schemes with guaranteed stability. In R. Berber & C. Kravaris (Eds.), *NATO ASI on nonlinear model based process control* (pp. 465–494). Dordrecht: Kluwer Academic.
2. Antoniades, C., & Christofides, P. D. (1999). Feedback control of nonlinear differential difference equation systems. *Chemical Engineering Science, 54*, 5677–5709.
3. Azimi-Sadjadi, B. (2003). Stability of networked control systems in the presence of packet losses. In *Proceedings of the 42nd IEEE conference on decision and control* (pp. 676–681). Maui, Hawaii.
4. Bemporad, A., & Morari, M. (1999). Control of systems integrating logic, dynamics and constraints. *Automatica, 35*, 407–427.
5. Bertsekas, D. P., & Tsitsiklis, J. N. (1997). *Parallel and distributed computation*. Belmont: Athena Scientific.
6. Bitmead, R. R., Gevers, M., & Wertz, V. (1990). *Adaptive optimal control—the thinking man's GPC*. Englewood Cliffs: Prentice-Hall.
7. Brockett, R. W., & Liberzon, D. (2000). Quantized feedback stabilization of linear systems. *IEEE Transactions on Automatic Control, 45*, 1279–1289.
8. Camponogara, E., Jia, D., Krogh, B. H., & Talukdar, S. (2002). Distributed model predictive control. *IEEE Control Systems Magazine, 22*, 44–52.
9. Chen, X., Liu, J., Muñoz de la Peña, D., & Christofides, P. D. (2010). Sequential and iterative distributed model predictive control of nonlinear process systems subject to asynchronous measurements. In *Proceedings of the 9th international symposium on dynamics and control of process systems* (pp. 611–616). Leuven, Belgium.
10. Chilin, D., Liu, J., Muñoz de la Peña, D., Christofides, P. D., & Davis, J. F. (2010). Detection, isolation and handling of actuator faults in distributed model predictive control systems. *Journal of Process Control, 20*, 1059–1075.
11. Christofides, P. D., & El-Farra, N. H. (2005). *Control of nonlinear and hybrid process systems: Designs for uncertainty, constraints and time-delays*. Berlin: Springer.
12. Christofides, P. D., Davis, J. F., El-Farra, N. H., Clark, D., Harris, K. R. D., & Gipson, J. N. (2007). Smart plant operations: vision, progress and challenges. *AIChE Journal, 53*, 2734–2741.
13. Cinar, A., Palazoglu, A., & Kayihan, F. (2007). *Chemical process performance evaluation*. Boca Raton: CRC Press/Taylor & Francis.
14. Clarke, F., Ledyaev, Y., & Sontag, E. (1997). Asymptotic controllability implies feedback stabilization. *IEEE Transactions on Automatic Control, 42*, 1394–1407.
15. Cover, T. M., & Thomas, J. A. (2002). *Elements of information theory*. New York: Wiley-Interscience.

P.D. Christofides et al., *Networked and Distributed Predictive Control,*
Advances in Industrial Control,
DOI 10.1007/978-0-85729-582-8, © Springer-Verlag London Limited 2011

16. Davis, J. F. (2007). *Report from NSF workshop on cyberinfrastructure in chemical and biological systems: impact and directions* (Technical report). (See http://www.oit.ucla.edu/nsfci/NSFCIFullReport.pdf for the pdf file of this report.)

17. Dunbar, W. B. (2007). Distributed receding horizon control of dynamically coupled nonlinear systems. *IEEE Transactions on Automatic Control, 52,* 1249–1263.

18. El-Farra, N. H., & Christofides, P. D. (2001). Integrating robustness, optimality and constraints in control of nonlinear processes. *Chemical Engineering Science, 56,* 1841–1868.

19. El-Farra, N. H., & Christofides, P. D. (2003). Bounded robust control of constrained multivariable nonlinear processes. *Chemical Engineering Science, 58,* 3025–3047.

20. Elia, N., & Eisenbeis, J. N. (2004). Limitations of linear remote control over packet drop networks. In *Proceedings of IEEE conference on decision and control* (pp. 5152–5157). Nassau, Bahamas.

21. Fogler, H. S. (1999). *Elements of chemical reaction engineering.* Englewood Cliffs: Prentice Hall.

22. Franco, E., Magni, L., Parisini, T., Polycarpou, M. M., & Raimondo, D. M. (2008). Cooperative constrained control of distributed agents with nonlinear dynamics and delayed information exchange: a stabilizing receding-horizon approach. *IEEE Transactions on Automatic Control, 53,* 324–338.

23. Ganji, H., Ahari, J. S., Farshi, A., & Kakavand, M. (2004). Modelling and simulation of benzene alkylation process reactors for production of ethylbenzene. *Petroleum and Coal, 46,* 55–63.

24. Gao, H., Chen, T., & Lam, J. (2008). A new delay system approach to network-based control. *Automatica, 44,* 39–52.

25. García, C. E., Prett, D. M., & Morari, M. (1989). Model predictive control: theory and practice—a survey. *Automatica, 25,* 335–348.

26. Ghantasala, S., & El-Farra, N. H. (2009). Robust diagnosis and fault-tolerant control of distributed processes over communication networks. *International Journal of Adaptive Control and Signal Processing, 23,* 699–721.

27. Grüne, L., Pannek, J., & Worthmann, K. (2009). A networked unconstrained nonlinear MPC scheme. In *Proceedings of ECC 2009* (pp. 371–376). Budapest, Hungary.

28. Hadjicostis, C. N., & Touri, R. (2002). Feedback control utilizing packet dropping networks links. In *Proceedings of the 41st IEEE conference on decision and control* (pp. 1205–1210). Las Vegas, Nevada.

29. Hassibi, A., Boyd, S. P., & How, J. P. (1999). Control of asynchronous dynamical systems with rate constraints on events. In *Proceedings of IEEE conference on decision and control* (pp. 1345–1351). Phoenix, Arizona.

30. Heidarinejad, M., Liu, J., Muñoz de la Peña, D., Davis, J. F., & Christofides, P. D. (2011). Handling communication disruptions in distributed model predictive control. *Journal of Process Control, 21,* 173–181.

31. Heidarinejad, M., Liu, J., Muñoz de la Peña, D., Christofides, P. D., & Davis, J. F. (2011, submitted). Multirate Lyapunov-based distributed model predictive control of nonlinear uncertain systems. *Automatica.*

32. Hespanha, J. P. (2005). A model for stochastic hybrid systems with application to communication networks. *Nonlinear Analysis, 62,* 1353–1383.

33. Hofierka, J., & Suri, M. (2002). The solar radiation model for open source GIS: implementation and applications. In *Proceedings of the open source GIS-GRASS users conference* (pp. 1–19). Trento, Italy.

34. Hong, S. H. (1995). Scheduling algorithm of data sampling times in the integrated communication and control-systems. *IEEE Transactions on Control Systems Technology, 3,* 225–230.

35. Imer, O. C., Yüksel, S., & Başar, T. (2006). Optimal control of LTI systems over unreliable communications links. *Automatica, 42,* 1429–1439.

36. Jeong, S. C., & Park, P. (2005). Constrained MPC algorithm for uncertain time-varying systems with state-delay. *IEEE Transactions on Automatic Control, 50,* 257–263.

37. Jia, D., & Krogh, B. (2002). Min-max feedback model predictive control for distributed control with communication. In *Proceedings of the American control conference* (pp. 4507–4512). Anchorage, Alaska.
38. Jogwar, S. S., Baldea, M., & Daoutidis, P. (2009). Dynamics and control of process networks with large energy recycle. *Industrial & Engineering Chemistry Research, 48*, 6087–6097.
39. Keviczky, T., Borrelli, F., & Balas, G. J. (2006). Decentralized receding horizon control for large scale dynamically decoupled systems. *Automatica, 42*, 2105–2115.
40. Khalil, H. K. (1996). *Nonlinear systems* (2nd ed.). New York: Prentice Hall.
41. Kokotovic, P., & Arcak, M. (2001). Constructive nonlinear control: a historical perspective. *Automatica, 37*, 637–662.
42. Kothare, S. L. D., & Morari, M. (2000). Contractive model predictive control for constrained nonlinear systems. *IEEE Transactions on Automatic Control, 45*, 1053–1071.
43. Kumar, A., & Daoutidis, P. (2002). Nonlinear dynamics and control of process systems with recycle. *Journal of Process Control, 12*, 475–484.
44. Lee, W. J. (2005). *Ethylbenzene dehydrogenation into styrene: kinetic modeling and reactor simulation*. PhD thesis, Texas A&M University, College Station, TX, USA.
45. Lian, F.-L., Moyne, J., & Tilbury, D. (2003). Modelling and optimal controller design of networked control systems with multiple delays. *International Journal of Control, 76*, 591–606.
46. Lin, H., & Antsaklis, P. J. (2005). Stability and persistent disturbance attenuation properties for a class of networked control systems: switched system approach. *International Journal of Control, 78*, 1447–1458.
47. Lin, Y., & Sontag, E. D. (1991). A universal formula for stabilization with bounded controls. *Systems & Control Letters, 16*, 393–397.
48. Lin, Y., Sontag, E. D., & Wang, Y. (1996). A smooth converse Lyapunov theorem for robust stability. *SIAM Journal on Control and Optimization, 34*, 124–160.
49. Liu, G.-P., Xia, Y., Chen, J., Rees, D., & Hu, W. (2007). Networked predictive control of systems with random networked delays in both forward and feedback channels. *IEEE Transactions on Industrial Electronics, 54*, 1282–1297.
50. Liu, J., Muñoz de la Peña, D., Christofides, P. D., & Davis, J. F. (2008a). Lyapunov-based model predictive control of particulate processes subject to asynchronous measurements. *Particle & Particle Systems Characterization, 25*, 360–375.
51. Liu, J., Muñoz de la Peña, D., Ohran, B. J., Christofides, P. D., & Davis, J. F. (2008b). A two-tier architecture for networked process control. *Chemical Engineering Science, 63*, 5394–5409.
52. Liu, J., Muñoz de la Peña, D., & Christofides, P. D. (2009). Distributed model predictive control of nonlinear process systems. *AIChE Journal, 55*, 1171–1184.
53. Liu, J., Muñoz de la Peña, D., Christofides, P. D., & Davis, J. F. (2009). Lyapunov-based model predictive control of nonlinear systems subject to time-varying measurement delays. *International Journal of Adaptive Control and Signal Processing, 23*, 788–807.
54. Liu, J., Chen, X., Muñoz de la Peña, D., & Christofides, P. D. (2010a). Iterative distributed model predictive control of nonlinear systems: handling delayed measurements. In *Proceedings of the 49th IEEE conference on decision and control* (pp. 7251–7258). Atlanta, Georgia.
55. Liu, J., Chen, X., Muñoz de la Peña, D., & Christofides, P. D. (2010b). Sequential and iterative architectures for distributed model predictive control of nonlinear process systems. *AIChE Journal, 56*, 2137–2149.
56. Liu, J., Muñoz de la Peña, D., & Christofides, P. D. (2010). Distributed model predictive control of nonlinear systems subject to asynchronous and delayed measurements. *Automatica, 46*, 52–61.
57. Liu, J., Muñoz de la Peña, D., Ohran, B. J., Christofides, P. D., & Davis, J. F. (2010c). A two-tier control architecture for nonlinear process systems with continuous/asynchronous feedback. *International Journal of Control, 83*, 257–272.
58. Liu, J., Chen, X., Muñoz de la Peña, D., & Christofides, P. D. (2011, submitted). Iterative distributed model predictive control of nonlinear systems: handling asynchronous, delayed measurements. *IEEE Transactions on Automatic Control*.

59. Maeder, U., Cagienard, R., & Morari, M. (2007). Explicit model predictive control. In S. Tarbouriech, G. Garcia, & A. H. Glattfelder (Eds.), *Lecture notes in control and information sciences: Vol. 346. Advanced strategies in control systems with input and output constraints* (pp. 237–271). Berlin: Springer.

60. Maestre, J. M., Muñoz de la Peña, D., & Camacho, E. F. (2009). A distributed MPC scheme with low communication requirements. In *Proceedings of the American control conference* (pp. 2797–2802). Saint Louis, MO, USA.

61. Maestre, J. M., Muñoz de la Peña, D., & Camacho, E. F. (2011). Distributed model predictive control based on a cooperative game. *Optimal Control Applications and Methods, 32*, 153–176.

62. Magni, L., & Scattolini, R. (2006). Stabilizing decentralized model predictive control of nonlinear systems. *Automatica, 42*, 1231–1236.

63. Mao, X. (1999). Stability of stochastic differential equations with Markovian switching. *Stochastic Processes and Their Applications, 79*, 45–67.

64. Massera, J. L. (1956). Contributions to stability theory. *Annals of Mathematics, 64*, 182–206.

65. Mayne, D. Q., Rawlings, J. B., Rao, C. V., & Scokaert, P. O. M. (2000). Constrained model predictive control: stability and optimality. *Automatica, 36*, 789–814.

66. McKeon-Slattery, M. (2010). The world of wireless. *Chemical Engineering Progress, 106*, 6–11.

67. Mhaskar, P., El-Farra, N. H., & Christofides, P. D. (2005). Predictive control of switched non-linear systems with scheduled mode transitions. *IEEE Transactions on Automatic Control, 50*, 1670–1680.

68. Mhaskar, P., El-Farra, N. H., & Christofides, P. D. (2006). Stabilization of nonlinear systems with state and control constraints using Lyapunov-based predictive control. *Systems & Control Letters, 55*, 650–659.

69. Mhaskar, P., Gani, A., McFall, C., Christofides, P. D., & Davis, J. F. (2007). Fault-tolerant control of nonlinear process systems subject to sensor faults. *AIChE Journal, 53*, 654–668.

70. Montestruque, L. A., & Antsaklis, P. J. (2003). On the model-based control of networked systems. *Automatica, 39*, 1837–1843.

71. Montestruque, L. A., & Antsaklis, P. J. (2004). Stability of model-based networked control systems with time-varying transmission times. *IEEE Transactions on Automatic Control, 49*, 1562–1572.

72. Muñoz de la Peña, D., & Christofides, P. D. (2008). Lyapunov-based model predictive control of nonlinear systems subject to data losses. *IEEE Transactions on Automatic Control, 53*, 2076–2089.

73. Muñoz de la Peña, D., & Christofides, P. D. (2008). Stability of nonlinear asynchronous systems. *Systems & Control Letters, 57*, 465–473.

74. Naghshtabrizi, P., & Hespanha, J. (2005). Designing an observer-based controller for a network control system. In *Proceedings of the 44th IEEE conference on decision and control and the European control conference 2005* (pp. 848–853). Seville, Spain.

75. Naghshtabrizi, P., & Hespanha, J. (2006). Anticipative and non-anticipative controller design for network control systems. In *Lecture notes in control and information sciences: Vol. 331. Networked embedded sensing and control* (pp. 203–218).

76. Nair, G. N., & Evans, R. J. (2000). Stabilization with data-rate-limited feedback: tightest attainable bounds. *Systems & Control Letters, 41*, 49–56.

77. Nešić, D., & Teel, A. R. (2004a). Input-output stability properties of networked control systems. *IEEE Transactions on Automatic Control, 49*, 1650–1667.

78. Nešić, D., & Teel, A. R. (2004b). Input-to-state stability of networked control systems. *Automatica, 40*, 2121–2128.

79. Nešić, D., Teel, A., & Kokotovic, P. (1999). Sufficient conditions for stabilization of sampled-data nonlinear systems via discrete time approximations. *Systems & Control Letters, 38*, 259–270.

80. Neumann, P. (2007). Communication in industrial automation: what is going on? *Control Engineering Practice, 15*, 1332–1347.

81. Nguyen, G. T., Katz, R. H., Noble, B., & Satyanarayananm, M. (1996). A tracebased approach for modeling wireless channel behavior. In *Proceedings of the winter simulation conference* (pp. 597–604). Coronado, California.
82. Peinke, J., Anahua, E., Barth, S., Goniter, H., Schaffarczyk, A. P., Kleinhans, D., & Friedrich, R. (2008). Turbulence a challenging issue for the wind energy conversion. In *Proceedings of 2008 European wind energy conference & exhibition*. Brussels Expo, Belgium.
83. Perego, C., & Ingallina, P. (2004). Combining alkylation and transalkylation for alkylaromatic production. *Green Chemistry, 6, 274–279*
84. Perk, S., Teymour, F., & Cinar, A. (2010). Statistical monitoring of complex chemical processes using agent-based systems. *Industrial & Engineering Chemistry Research, 49,* 5080–5093.
85. Ploplys, N. J., Kawka, P. A., & Alleyne, A. G. (2004). Closed-loop control over wireless networks—developing a novel timing scheme for real-time control systems. *IEEE Control Systems Magazine, 24,* 52–71.
86. Primbs, J. A., Nevistic, V., & Doyle, J. C. (2000). A receding horizon generalization of pointwise min-norm controllers. *IEEE Transactions on Automatic Control, 45,* 898–909.
87. Proakis, J. G., & Salehi, M. (2007). *Digital communications* (5th ed.). Columbus: McGraw-Hill.
88. Qi, W., Liu, J., Chen, X., & Christofides, P. D. (2011). Supervisory predictive control of stand-alone wind–solar energy generation systems. *IEEE Transactions on Control Systems Technology, 19,* 199–207.
89. Qin, S. J., & Badgwell, T. A. (2003). A survey of industrial model predictive control technology. *Control Engineering Practice, 11,* 733–764.
90. Raimondo, D. M., Magni, L., & Scattolini, R. (2007). Decentralized MPC of nonlinear system: an input-to-state stability approach. *International Journal of Robust and Nonlinear Control, 17,* 1651–1667.
91. Rawlings, J. B. (2000). Tutorial overview of model predictive control. *IEEE Control Systems Magazine, 20,* 38–52.
92. Rawlings, J. B., & Stewart, B. T. (2008). Coordinating multiple optimization-based controllers: New opportunities and challenges. *Journal of Process Control, 18,* 839–845.
93. Richards, A., & How, J. P. (2007). Robust distributed model predictive control. *International Journal of Control, 80,* 1517–1531.
94. Ritchey, V. S., & Franklin, G. F. (1989). A stability criterion for asynchronous multirate linear systems. *IEEE Transactions on Automatic Control, 34,* 529–535.
95. Scattolini, R. (2009). Architectures for distributed and hierarchical model predictive control—a review. *Journal of Process Control, 19,* 723–731.
96. Shin, K. G. (1991). Real-time communications in a computer-controlled workcell. *IEEE Transactions on Robotics and Automation, 7,* 105–113.
97. Sontag, E. (1989). A 'universal' construction of Artstein's theorem on nonlinear stabilization. *Systems & Control Letters, 13,* 117–123.
98. Stewart, B. T., Venkat, A. N., Rawlings, J. B., Wright, S. J., & Pannocchia, G. (2010). Cooperative distributed model predictive control. *Systems & Control Letters, 59,* 460–469.
99. Su, Y. F., Bhaya, A., Kaszkurewicz, E., & Kozyakin, V. S. (1997). Further results on stability of asynchronous discrete-time linear systems. In *Proceedings of the 36th IEEE conference on decision and control* (pp. 915–920). San Diego, California.
100. Sun, Y., & El-Farra, N. H. (2008). Quasi-decentralized model-based networked control of process systems. *Computers & Chemical Engineering, 32,* 2016–2029.
101. Tabbara, M., Nešić, D., & Teel, A. R. (2007). Stability of wireless and wireline networked control systems. *IEEE Transactions on Automatic Control, 52,* 1615–1630.
102. Tatara, E., Cinar, A., & Teymour, F. (2007). Control of complex distributed systems with distributed intelligent agents. *Journal of Process Control, 17,* 415–427.
103. Tipsuwan, Y., & Chow, M. (2003). Control methodologies in networked control systems. *Control Engineering Practice, 11,* 1099–1111.

104. Valenciaga, F., & Puleston, P. F. (2005). Supervisor control for a stand-alone hybrid genera-
     tion system using wind and photovoltaic energy. *IEEE Transactions on Energy Conversion*,
     20, 398–405.
105. Valenciaga, F., Puleston, P. F., Battaiotto, P. E., & Mantz, R. J. (2000). Passivity/sliding mode
     control of a stand-alone hybrid generation system. *IEE Proceedings. Control Theory and
     Applications*, *147*, 680–686.
106. Valenciaga, F., Puleston, P. F., & Battaiotto, P. E. (2001). Power control of a photovoltaic
     array in a hybrid electric generation system using sliding mode techniques. *IEE Proceedings.
     Control Theory and Applications*, *148*, 448–455.
107. Valenciaga, F., Puleston, P. F., & Battaiotto, P. E. (2004). Variable structure system control
     design method based on a differential geometric approach: application to a wind energy
     conversion subsystem. *IEE Proceedings. Control Theory and Applications*, *151*, 6–12.
108. Venkat, A. N., Rawlings, J. B., & Wright, S. J. (2005). Stability and optimality of distributed
     model predictive control. In *Proceedings of the 44th IEEE conference on decision and control
     and the European control conference ECC 2005* (pp. 6680–6685). Seville, Spain.
109. Wächter, A., & Biegler, L. T. (2006). On the implementation of primal-dual interior point
     filter line search algorithm for large-scale nonlinear programming. *Mathematical Program-
     ming*, *106*, 25–57.
110. Walsh, G., Beldiman, O., & Bushnell, L. (2001). Asymptotic behavior of nonlinear net-
     worked control systems. *IEEE Transactions on Automatic Control*, *46*, 1093–1097.
111. Walsh, G., Ye, H., & Bushnell, L. (2002). Stability analysis of networked control systems.
     IEEE Transactions on Control Systems Technology, *10*, 438–446.
112. Wang, Y. M. L., Chu, T., & Hao, F. (2005). Stabilization of networked control systems with
     data packet dropout and transmission delays: continuous-time case. *European Journal of
     Control*, *11*, 40–49, 55.
113. Witrant, E., Georges, D., Canudas-de-Wit, C., & Alamir, M. (2007). On the use of state
     predictors in networked control system. In *Lecture notes in control and information sciences:
     Vol. 352. Applications of time delay systems* (pp. 17–35). New York: Springer.
114. Ydstie, E. B. (2002). New vistas for process control: Integrating physics and communication
     networks. *AIChE Journal*, *48*, 422–426.
115. Ye, H., & Walsh, G. (2001). Real-time mixed-traffic wireless networks. *IEEE Transactions
     on Industrial Electronics*, *48*, 883–890.
116. Ye, H., Walsh, G., & Bushnell, L. (2000). Wireless local area networks in the manufactur-
     ing industry. In *Proceedings of the American control conference* (pp. 2363–2367). Chicago,
     Illinois.
117. You, H., Long, W., & Pan, Y. (2006). The mechanism and kinetics for the alkylation of
     benzene with ethylene. *Petroleum Science and Technology*, *24*, 1079–1088.
118. Zhang, L., Shi, Y., Chen, T., & Huang, B. (2005). A new method for stabilization of net-
     worked control systems with random delays. *IEEE Transactions on Automatic Control*, *50*,
     1177–1181.
119. Zornio, P., & Karschnia, B. (2009). Realizing the promise of wireless. *Chemical Engineering
     Progress*, *105*, 22–29.

Index

P.D. Christofides et al., *Networked and Distributed Predictive Control,*
Advances in Industrial Control,
DOI 10.1007/978-0-85729-582-8, © Springer-Verlag London Limited 2011

Other titles published in this series (continued):